Climate Change, Carbon Trading and Civil Society

Climate Change, Carbon Trading and Civil Society

Negative Returns on South African Investments

Edited by

Patrick Bond, Rehana Dada and Graham Erion

Published in 2009 by University of KwaZulu-Natal Press
Private Bag X01
Scottsville 3209
South Africa
Email: books@ukzn.ac.za
Website: www.ukznpress.co.za

First edition published by Rozenberg Publishers, The Netherlands
Second edition published by University of KwaZulu-Natal Press 2007

ISBN 978 1 86914 141 7

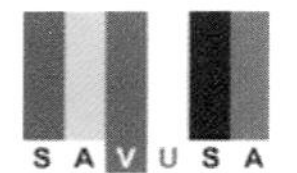

SAVUSA (South Africa – VU University – Strategic Alliances) co-ordinated the early stages of the preparation of this manuscript for publication.

Printed and bound by Pinetown Printers

Contents

Dedicated to the Memory of Sajida Khan

On 15 July 2007, Sajida Khan, aged 55, died at home of cancer caused – she was convinced – by Durban's largest dump, located across from her family residence in 1980 by the apartheid regime. A large number of her neighbours also succumbed to cancers, she documented. As the research director of the Cancer Association of SA once remarked, 'Clare Estate residents are like animals involved in a biological experiment.'

Passionate to a fault, Khan was self-taught and supremely confident when testifying about chemical pollution and the economics of solid waste. She earned a bachelors degree in microbiology at the former University of Durban-Westville, began work at Unilever, and soon invented a freeze-dried food formula that was patented. But she deregistered the patent so as to make it more accessible for low-income people across the world.

Khan became an activist because, as she observed, 'As early as 1987 the city promised to close this dump site and in its place give us all these sports fields. And they broke that promise to us. And again, for the 1994 election, the political parties also promised to close the dump, decommission it, and relocate the Clare Estate dump site. Again they broke that promise to us. Before the permit was granted, they should have created a buffer zone of 800 metres minimum to protect the people and that wasn't done.'

Khan was renowned for her hospitality, and visiting environmentalists made a pilgrimage to Bisasar Road, ranking it high amongst Durban's numerous 'toxic tour' sites. Inside were her generosity, fine refreshments and doctoral-level lectures in plant ecology and public health. Outside, a few dozen metres away, was Africa's first pilot project in carbon trading, in which methane from rotting trash will be extracted and the greenhouse gas reduction credits sold to Northern investors, in a plan initially endorsed by the World Bank. It is an innovation that municipal officials brag about – but that also stalled the dump's closure.

After Khan filed an Environmental Impact Assessment challenge, the Bank backed off, a victory that helped raised the profile of numerous other carbon offset problems (although the Bank funded two similar but much smaller landfill projects in Durban in mid-2007).

An international network against carbon trading, the Durban Group for Climate Justice, was founded in 2004 in part because of her charisma. According to Javier Baltodano and Isaac Rojas of Friends of the Earth, Costa Rico, 'Sajida introduced us to how carbon credits were used to justify the dump in the middle of a neighbourhood. She showed us her strong willingness to resist over the sickness, the dump and the racism.'

Says Durban environmentalist Muna Lakhani, 'We have lost a sister, a stalwart, a spirit that I have known well for over 30 years. I miss her, but am glad that her suffering is over. Please can we choose to live our lives just a little bit in her memory, so that our consumption of our planet's resources does not lead to more Sajidas?'

Khan is survived by her mother Kathija and siblings Hanifa, Zainuladevien, Rafique and Akram.

Patrick Bond

(originally published in Mail & Guardian, *23 July 2007)*

Abbreviations

ACP	Africa-Caribbean-Pacific
ACUNS	Academic Council for the United Nations System
AEC	Atomic Energy Corporation
AGOA	Africa Growth and Opportunity Act
ANC	African National Congress
ATTAC	Association for the Taxation of Financial Transactions for the Aid of Citizens
BASE	Basel Agency for Sustainable Energy
BBC	British Broadcasting Corporation
BEE	black economic empowerment
CAN	Climate Action Network
CAS	country assistance strategy
CAT	Centre for Alternative Technology
CCS	Centre for Civil Society
CDCF	Community Development Carbon Fund
CDF	Clean Development Fund
CDM	Clean Development Mechanism
CEF	Central Energy Fund
CEO	chief executive officer
CEO	Corporate Europe Observatory
CER	Certified Emissions Reduction
CFL	compact fluorescent light
COP	Conference of Parties
COSATU	Congress of South African Trade Unions
DEAT	Department of Environmental Affairs and Tourism
DI	direct investment
DNA	Designated National Authority
DNV	Det Norske Veritas
DOE	Designated Operational Entity
DRC	Democratic Republic of the Congo

EDRC	Energy for Development Research Centre
EIA	environmental impact assessment
EIR	Extractive Industries Review
EMS	Environmental Media Services
ENDS	Environmental Data Services
ENGO	environmental non-governmental organisation
EPA	economic partnership agreement
ERA	Environmental Rights Action
EROI	energy return on energy investment
ETS	Emissions Trading Scheme
EU	European Union
FAO	Food and Agriculture Organisation
FASE	Foundation for Advancement in Science and Education
FCPF	Forest Carbon Partnership Facility
FDI	foreign direct investment
FSC	Forest Stewardship Council
GDP	gross domestic product
GE	genetically engineered
GEF	Global Environmental Facility
GNI	gross national income
GNP	gross national product
GNS	gross national saving
HSRC	Human Sciences Research Council
HRW	Human Rights Watch
ICC	International Chamber of Commerce
IDC	Industrial Development Corporation
IEA	International Energy Agency
IETA	International Emissions Trading Association
IFC	International Finance Corporation
IGCC	integrated gasification combine cycle
IMF	International Monetary Fund
INC	International Negotiating Committee
IPCC	Intergovernmental Panel on Climate Change
IRIN	Integrated Regional Information Network
IRR	internal rate of return
ITT	Ishpingo-Tambococha-Tiputini

IUCN	International Union for the Conservation of Nature (World Conservation Union)
JI	joint implementation
LDC	less-developed country
MCA	Millenium Challenge Account
MDB	multilateral development bank
MOSOP	Movement for the Survival of the Ogoni People
NASA	National Aeronautics and Space Administration
NATO	North Atlantic Treaty Organisation
NECSA	Nuclear Energy Corporation of South Africa
NEPAD	New Partnership for Africa's Development
NERSA	National Energy Regulator of South Africa
NGO	non-governmental organisation
NNPC	Nigerian National Petroleum Company
NNR	National Nuclear Regulator
ODA	Overseas Development Aid
OPEC	Organisation of Petroleum Exporting Countries
PBMR	pebble bed modular reactor
PCF	Prototype Carbon Fund
PIN	Project Identification Note
PPP	public-private partnership
PV	photovoltaic
RECLAIM	Regional Clean Air Incentives Market
REDD	Reducing Emissions from Deforestation and Degradation
RIVM	The National Institute of Public Health
ROT	rehabilitate, operate and transfer
SABC	South African Broadcasting Corporation
SACAN	South African Climate Action Network
SANPAD	South Africa-Netherlands Research Programme on Alternatives in Development
SAP	structural adjustment policy
SAPA	South African Press Association
SEEN	Sustainable Energy and Economy Network
SSN	SouthSouthNorth
TNI	Transnational Institute
UN	United Nations
UNCTAD	United Nations Conference on Trade and Development

UNDP	United Nations Development Programme
UNFCCC	United Nations Framework Convention on Climate Change
US	United States of America
WBCSD	World Business Council for Sustainable Development
WBG	World Bank Group
WRM	World Rainforest Movement
WTO	World Trade Organisation
WWF	Worldwide Fund for Nature

Acknowledgements

This book was made possible not only because of dedicated CCS/TNI researchers and a very capable team at Rozenberg Publishers in Amsterdam along with University of KwaZulu-Natal Press, but also because of our far-sighted financial sponsors, the SA-Netherlands Research Programme on Alternatives in Development. SANPAD pursues the following objectives, with which we agree entirely: 'To stimulate and promote quality research; to produce research outputs intended and useful for development purposes; to promote co-operation between Dutch and South African researchers, and between institutions within South Africa; and to develop research capacity and a culture conducive to research, aimed particularly at researchers from historically disadvantaged communities'.

SANPAD director Anshu Padayachee is especially thanked for her support, as are the Transnational Institute's CarbonTradeWatch, TNI director Fiona Dove, and TNI public services/energy specialist Daniel Chavez, who assisted us in February 2004 with project design. In June and October 2005, and July 2006, SANPAD and TNI supported Centre for Civil Society colloquia on energy and climate change, the latter arranged by the Centre for Civil Society with assistance from Durban-based Timberwatch. We also appreciate the support of the Netherlands Institute for Southern Africa in facilitating contact with the Amsterdam team.

Naturally we also thank our contributors to this volume, including some from institutions with which we have strong relations, such as the Dag Hammarskjöld Foundation (whose carbon trading seminars in Uppsala in September 2006 and at the Nairobi World Social Forum in January 2007 were invaluable at the late stages of this book's compilation), The Corner House, SinksWatch and the Sustainable Energy and Economy Network at the Institute for Policy Studies.

An earlier version of this book – *Trouble in the Air* – was published as a 'civil society reader' by TNI and CCS in October 2004, and contained journalistic articles and reports from sites of struggle, which are posted at the CCS and Carbon Trade Watch websites. We thank the additional contributors to that volume, including Janet Wilhelm, Shankar Vedantam, Megan Lindow, Richard Worthington, Juggie Naran and especially Mpumelelo Mhlalisi and Caroline Ntaopane, who maintain a vigilant grass-roots watch on CDM projects in Cape Town and the Steel Valley.

Additionally, we have used graphics and data originally assembled by, amongst others, Mark Jury of the University of Zululand and Anton Eberhardt of the University of Cape Town, as well as by the environmental economics staff at the World Bank (an institution whose policies have been profoundly damaging in this and so many other areas). We're very grateful for the hard work that went into these.

Mostly, we're grateful to our colleagues in the Durban Group, who from October 2004 have been encouraging our efforts, and many others' struggles for climate justice across the world.

Patrick Bond, Rehana Dada and Graham Erion

Contributors

Heidi Bachram is a member of the Amsterdam-based Transnational Institute's Carbon Trade Watch collective and also works for the Africa information network Fahamu in Oxford.

Vanessa Black is a green architect and an activist with Earthlife Africa.

Patrick Bond is director of the Centre for Civil Society and professor of Development Studies at the University of KwaZulu-Natal.

Rehana Dada is an environmental journalist and a postgraduate student at the Centre for Civil Society, University of KwaZulu-Natal.

Michael K. Dorsey is assistant professor of Environment at Dartmouth College, USA, and a leading ecological anti-racism activist.

Graham Erion is based at the York University School of Law and Faculty of Environmental Studies in Toronto. He is a Transnational Institute Carbon Trade Watch research associate, and served as a Centre for Civil Society visiting scholar and trainer in mid-2005.

groundWork is an award-winning Pietermaritzburg-based environmental justice non-governmental organisation, whose studies include *The groundWork Report 2005: Whose energy future? Big oil against people in Africa.*

Jutta Kill heads up SinksWatch and is a founding member of the Durban Group for Climate Justice.

Muna Lakhani is a Centre for Civil Society energy research associate, an activist with Earthlife Africa, and founder of the Institute for Zero Waste in Africa.

Larry Lohmann works at The Corner House, a British eco-social think tank, and recently edited a special issue of *Development Dialogue* on carbon trading.

Joan Martinez-Alier is professor at the Universitat Autonoma de Barcelona, a past president of the International Society for Ecological Economics, and honorary rector of the Instituto de Estudios Ecologitas del Tercer Mundo in Quito, Equador.

Trusha Reddy was a Centre for Civil Society visiting scholar in early 2005, subsequently pursued postgraduate studies at the New School for Social Research in New York City, and currently works at the Institute of Security Studies.

Janet Redman is a researcher with the Sustainable Energy and Economy Network at the Institute for Policy Studies in Washington, DC.

Leah Temper is a doctoral student in ecological economics at the Universitat Autonoma de Barcelona.

Brian Tokar is director of the Institute for Social Ecology in Plainfield, Vermont, USA.

Daphne Wysham is the founder and co-director of the Sustainable Economy and Energy Network and a fellow at the Institute for Policy Studies in Washington, DC.

Introduction

Patrick Bond, Rehana Dada and Graham Erion

A brief history of carbon trading

The intellectual origins of carbon trading can be traced back to a small publication in 1968 titled 'Pollution, Property, and Prices' by Canadian economist John Dales. Like Garrett Hardin – who penned his famous essay, 'The Tragedy of the Common' in the same year – Dales believed that natural resources in their unrestricted common property form would face tragic exploitation by people acting in their rational self-interest (Hardin 1968). Yet Dales went much further than Hardin in his solution to the problem. Dales proposed to control water pollution by setting a total quota of allowable waste for each waterway and then setting up a 'market' in equivalent 'pollution rights' to firms to discharge pollutants up to this level (Dales 1968: 81). These rights, referred to as 'transferable property rights . . . for the disposal of wastes' would be sold to firms, which could then trade them among themselves (Dales 1968: 85). The more efficient firms would make the largest pollution reductions and sell their credits to less efficient firms, thereby guaranteeing a reduction of pollution at the lowest social cost.

Although Dales' proposal took a backseat to the command-and-control approach that characterised environmental policy during the 1970s, his idea would resurface in the decades that followed. Proponents of pollution trading – typically a mix of industry groups and self-described 'free-market environmentalists' – echoed Dales' logic about greater efficiency and added claims of lower administrative costs and greater incentives for innovation. After a series of proposals and pilot projects by the Environmental Protection Agency, the United States Congress amended the Clean Air Act in 1990 to create a national emissions-trading scheme in sulphur dioxide, the main pollutant behind acid rain. Until 1997, the United States was the only country in the world with any significant pollution trading scheme, but this would change significantly, following the Kyoto Protocol.

Although carbon trading was initially met with hostility from some European countries and environmental non-governmental organisations (ENGOs) during the third Conference of Parties (COP) to the United Nations Framework Convention on Climate Change (UNFCCC) in Kyoto, it was eventually adopted and appears in three separate articles of the final text of the Protocol. Article 17 of the Protocol establishes a system of 'emissions trading' whereby Annex 1 countries (developed countries that have accepted binding emissions reductions targets) can trade emissions credits among themselves if they overshoot their targets. This aspect of trading can be controversial, especially when applied to former Eastern Bloc countries such as Russia and the Ukraine. The collapse of the Soviet economy during the 1990s meant that these countries today have a 'free pass' on reducing emissions. Their contracted economies have already reduced gross emissions by nearly 40 per cent (BBC 2005).

For this reason the unfavourable label 'hot air' has been widely applied to this form of trading, since it has nothing to do with deliberate efforts to reduce emissions and everything to do with economic collapse. However, Europe's larger considerations about energy security and access to natural gas may still benefit Russia and the Ukraine in this market and increase the likelihood of future trading under article 17 (PCF 2005a: 45).

The second type of carbon trading is joint implementation (JI) – article 4 – whereby Annex 1 countries can invest in projects in other Annex 1 countries, with the investing country receiving credit for the host country's reductions. Since in practice this type of trading has been limited to Eastern European countries and is not allowed within EU member countries, it has yet to gain the prominence of the Clean Development Mechanism (CDM). However, with a much more lenient scheme for project approvals and supervision, it is likely that more JI projects will be seen in the future.

With emissions trading and JI playing minimal roles, the global carbon market is at present almost entirely made up of transactions under article 12 of the Kyoto Protocol, the CDM. The CDM provides an opportunity for Annex 1 countries to receive emissions reduction credits to use against their own targets by investing in projects to reduce or sequester greenhouse gas emissions in non-Annex 1 countries (developing countries).

One of the most controversial aspects of article 12 is that it requires projects to show 'reductions in emissions that are additional to any that would occur in the absence of the certified project activity' (The Kyoto Protocol, article 12, paragraph 5.2). This requirement has become known as 'additionality' and is intended to ensure there is a net emissions reduction.

Another controversial aspect of the CDM is the requirement that projects must help developing countries in 'achieving sustainable development' (The Kyoto Protocol, article 12, paragraph 2). The sustainable development requirement represented a hard-fought victory by many of the countries and ENGOs that were initially against the CDM. However, in subsequent meetings of the COP, the UNFCCC countries have been allowed to set their own definition of sustainable development and judge whether a project meets these criteria, rather than adopting a universal definition that could better ensure the accountability of those authorities overseeing project approval.

A variety of domestic and international governance structures have been set up to oversee CDM projects. There are three key institutions governing CDM projects through their validation. The first of these is each host country's Designated National Authority (DNA). The DNA is the first institution to review a project's documents, namely the project design document that lays out all the relevant information about the project. As to the actual makeup of the DNAs, they will often be housed in government departments and staffed with public sector employees, such as in South Africa where the DNA is in the Department of Minerals and Energy. However, in other cases, such as Cambodia, the DNA is contracted out to private consultancies.

Assuming everything is in order, the DNA then writes a letter of approval, saying that all participants are voluntary and that the sustainable development criteria have been met. The project design document is then assessed by a Designated Operational Entity (DOE). Unlike the DNAs, the DOEs are all private sector entities. To date, twelve companies have been accredited as DOEs, although not all of them can assess every single methodology. To validate a project, the DOE will review the project design document to consider whether the project's methodology is in line with approved methodologies, if the claimed emissions reductions and baseline scenarios are accurate and to ensure that the project meets the condition of additionality. In making its determination, the DOE will post the project design document on the Internet for a 30-day public comment period.

With the approval of the DOE and the DNA, the final stage in project validation is the CDM executive board, where the findings of the DOE and DNA are reviewed and a final decision is made whether to allow the project to start generating Certified Emissions Reductions (CERs). There is also a final 30-day public comment period while the project is with the CDM executive board. With only twelve members, the board does not have the resources to closely scrutinise every project and as a result, relies heavily on the decisions of DOEs. According to Eric Haites, a private sector consultant in the carbon market, 'the vast majority of validation and certification decisions by DOEs are expected to be final; the executive board only deals with the problem cases' (Haites 2006).

Carbon market trends

With the process of validation now established and some of the relevant institutions explained, let us turn our attention to how the global carbon market has developed since

Kyoto. The first thing to note is the large role played by Northern firms and consultants – such as Ecosecurities – which are able to provide a certain level of capacity and expertise that might not be as readily accessible in Southern countries. Another example of this has been the prominence of the World Bank's Prototype Carbon Fund (PCF). In partnership with six governments and seventeen companies plus a budget of US$180 million, the PCF describes itself as 'a leader in the creation of a carbon market to help deal with the threat posed by climate change' (PCF 2004a: 7). As the single largest purchaser of CERs, as of September 2006, the PCF had 32 projects in development, with a total CER value potential of US$165 million.

A second noteworthy trend is that the market is heavily concentrated in large middle-income countries led by India, China and Brazil. The PCF admits that 'this concentration of CDM flows towards large middle-income countries is consistent with the current direction of foreign direct investment' (PCF 2004a: 5). By contrast poorer countries, especially in Africa, have been almost entirely left behind. As of September 2006, South Africa and Morocco were the only countries on the African continent to have validated a CDM project. According to the PCF, 'This under-representation of Africa raises deep concerns about the overall equity of the distribution of the CDM market, as the vast majority of African countries have not, for the moment, been able to pick up even one first deal' (PCF 2005a: 25). Early evidence hence contradicts the notion that the CDM will uplift the world's poorest countries to a cleaner path of development.

The other major trend in the carbon market has been the enormous profitability of non-carbon related projects. While renewable energy projects, which offset carbon dioxide (CO_2) emissions, make up nearly 58 per cent of the total number of projects, they account for only 15 per cent of the total number of CERs that have been issued (Fenhann 2006). By contrast, projects abating nitrous oxide and hydrofluorocarbons are less than 2 per cent of the overall *number* of projects, yet make up 74 per cent of the CERs issued to date by project sector (PCF 2005a).

These projects are known as 'low-hanging fruit', since their high returns mean they are the first to be picked by investors. The reason is that hydrofluorocarbon has 11 700 times the potency of CO_2 and since credits are in CO_2 equivalent (CO_2e), a relatively small capture of hydrofluorocarbon can bring an enormous windfall of credits. According to the PCF, the large amount of non-CO_2 projects in the carbon market has meant that 'traditional energy efficiency or fuel switching projects, which were initially expected to represent the bulk of the CDM, account for less than 5 per cent [of it now]' (PCF 2005a: 5). How these trends affect the legitimacy of the carbon market and to whose benefit are central questions addressed in this book.

The critique of carbon trading

To turn quickly to the main point we make in this book, what is the problem with carbon trading? Firstly, one has to consider the moral question. A commodity is being created: the property right to dirty the air and to heat the atmosphere. But who is in control of the process and who benefits from this commodity? Ironically, it is those who bear the historic legacy of emissions: the corporations and the World Bank are currently being rewarded with deeded rights to continue polluting.

What this means, in lay language, is, according to Daniel Becker of the Sierra Club's Global Warming and Energy Program, 'the moral equivalent of hiring a domestic. We will pay you to clean our mess. For a long time here in America we have believed in the polluter pays principle. This could become a pay to pollute principle' (cited in World Bank press clips, 29 November 2005).

The analogy is quite poignant. In South Africa, low wages and awful working conditions, including migrant relationships faced by domestic workers, reflect an historical legacy of injustice known as apartheid. The oppressive system was subsequently deracialised in formal legal terms, but simultaneously cemented by market-oriented labour relations since 1994. It is in this context that reparations were rejected as a strategy (or even demand) by the corporate-friendly ANC government and that the unemployment rate doubled. Like carbon trading, in which those who made the global ecological mess are now benefiting in the alleged clean-up strategy, South African social relations post-apartheid proved that it paid to pollute the society with racism, because most of the pre-1994 beneficiaries did not go to the Truth and Reconciliation Commission, nor have they lost their inherited wealth post-1994.

In the same way that apartheid represented a gift to white people, according to Larry Lohmann from the British non-governmental organisation (NGO), The Corner House and the Durban Group for Climate Justice (a network which formed in 2004 to oppose carbon trading), 'the distribution of carbon allowances (a prerequisite for trading) to the biggest polluters presupposes *one of the largest and most regressive schemes for creating property rights in history*' (2005; emphasis added).

What President Thabo Mbeki calls 'global apartheid' is hence reinforced by the Kyoto Protocol's carbon trading components, in part because it is based upon an historical allocation of pollution rights. To borrow Becker's metaphor, this occurs in the same way that pre-existing wage and social relations – evident especially in inexpensive domestic labour – are cemented by strengthened property rights in contemporary South Africa, in the wake of what the United Nations (UN) rightly termed a crime against humanity.

The second objection to the emissions trade is that the market is failing – even on its own narrow terms – as shown in great detail throughout this book. Because emissions cuts are not yet underway, carbon trading has become the key response of the international community – and South Africa – to the climate crisis, both in the form of emissions trading and in the form of trading in carbon credits. According to Lohmann,

> There is a critical distinction between pure emissions trading (for example, sulphur dioxide trading in the US, or the European Union's Emissions Trading System minus the linking directive, or the Kyoto Protocol not including Joint Implementation and the Clean Development Mechanism) on the one hand, and on the other, trading in credits from projects (the CDM, World Bank Prototype Carbon Fund, Carbon Neutral Company, etc.). Kyoto, the World Bank, and private corporations are constantly seeking to blur this distinction and tell us that by investing in windmills or light bulbs they are 'making emissions reductions' or doing something that is equivalent to emissions trading. It's not equivalent. (personal communication, 25 January 2005)

This market does not lack for controversy, particularly because fatuous carbon offset public relations schemes have persuaded politicians and celebrities that they can make their global conferences, rock concerts and other extravaganzas carbon neutral. As is increasingly clear, such offsets are often scams. *Guardian* columnist George Monbiot considered Schiermeier's 2006 *Nature* report on worsening global warming attributable to increased methane emissions from plants, which were formerly thought to be solely a sink for CO_2 emissions:

> While they have a pretty good idea of how much carbon our factories and planes and cars are releasing, scientists are much less certain about the amount of carbon tree planting will absorb. When you drain or clear the soil to plant trees, for example, you are likely to release some carbon, but it is hard to tell how much. Planting trees in one place might stunt trees elsewhere, as they could dry up a river which was feeding a forest downstream. Or by protecting your forest against loggers, you might be driving them into another forest.
>
> As global temperatures rise, trees in many places will begin to die back, releasing the carbon they contain. Forest fires could wipe them out completely. The timing is also critical: emissions saved today are far more valuable, in terms of reducing climate change, than emissions saved in ten years' time, yet the trees you plant start absorbing carbon long after your factories released it. All this made the figures speculative, but the new findings, with their massive uncertainty range (plants, the researchers say, produce somewhere between ten and thirty per cent of the planet's methane) make an honest sum impossible.

> In other words, you cannot reasonably claim to have swapped the carbon stored in oil or coal for carbon absorbed by trees. Mineral carbon, while it remains in the ground, is stable and quantifiable. Biological carbon is labile and uncertain. (2006)

As Monbiot concluded, 'perhaps the most destructive effect of the carbon offset trade is that it allows us to believe we can carry on polluting. The government can keep building roads and airports and we can keep flying to Thailand for our holidays, as long as we purchase absolution by giving a few quid to a tree planting company. How do you quantify complacency?'

In economic terms, can emissions trading work as designed, regardless of the ethical and ecological shortcomings? Climate activists such as the Durban Group for Climate Justice worry that there are very serious theoretical problems with the carbon market, including most variants of carbon taxation, which economists would recognise if they gave it serious thought. As Gar Lipow explains:

> Both emissions trading and green taxes are an inefficient way of reducing carbon emissions because they are largely driven by fossil fuel consumption and fossil fuel demand is extremely *price inelastic* (no matter how high the price goes, you are dependent and will find it hard to cut consumption).
>
> In the short run, some savings may be achieved by simple behaviour changes. But past a certain point you are giving up the ability to heat your home, get to work and in general experience other things vital to a decent life – so in the face of higher energy prices you will give up something else and simply pay for more energy.
>
> In the longer run, better capital investments can reduce such consumption without giving up vital things. But a combination of unequal access to capital, split incentives (where the person who makes the investment is not the one who would obtain the savings), transaction costs of energy savings vs. other investments, and other factors mean capital investment does not occur as you would expect in the face of rising energy prices. (2006)

Market players know the gimmicks; politicians promote the trade

What Monbiot and Lipow reveal above – namely, that the internal mechanisms of the carbon trade are fatally flawed – has become increasingly obvious to those active in the markets, or reporting on them for the financial press. Since the Kyoto Protocol's ratification on 16 February 2005, entrenching the nascent global emissions trade in international law, a series of scandals and market mishaps have emerged from files of dismayed financiers and business journalists.

The intrinsic problem in setting an artificially generated market price for carbon was revealed to the world in April 2006, when the European Union's emissions trading market crashed, thanks to the over-allocation of pollution rights. The carbon spot market price lost over half its value in a single day, destroying many CDM projects earlier considered viable investments. This was not an insignificant crash in financial terms, as the first quarter of 2006 had witnessed US$7.5 billion in carbon trades, as compared with US$11 billion (704 tonnes) in 2005, according to the World Bank.

Jutta Kill collected some examples of insider concern:

- As Tony Ward of Ernst & Young put it in May 2006, the prototype European Emissions Trading Scheme (ETS) 'has not encouraged meaningful investment in carbon-reducing technologies'.
- Added Peter Atherton of Citigroup in January 2007, 'ETS has done nothing to curb emissions . . . [and] is a highly regressive tax falling mostly on poor people'. The trade 'enhances the market power of generators'. Asking whether policy goals were achieved, he answered: 'Prices up, emissions up, profits up . . . so, not really'. Who wins and who loses? 'All generation-based utilities – winners. Coal and nuclear-based generators – biggest winners. Hedge funds and energy traders – even bigger winners. Losers . . . ahem . . . Consumers!'
- Speaking to Channel Four news in March 2007, the European commissioner for energy offered this verdict on the ETS: 'A failure'.
- The *Wall Street Journal* confirmed in March 2007 that emissions trading 'would make money for some very large corporations, but don't believe for a minute that this charade would do much about global warming'. The paper termed the carbon trade 'old-fashioned rent-seeking . . . making money by gaming the regulatory process'.
- What about carbon trading in the Third World, through the CDM? According to *Newsweek* magazine's investigation of carbon trading on 12 March 2007: 'It isn't working . . . [and represents] a grossly inefficient way of cutting emissions in the developing world'. The magazine called the trade 'a shell game' which has already transferred '$3 billion to some of the worst carbon polluters in the developing world'.
- Two weeks later, *Business Week* observed, 'Some deals amount to little more than feel-good hype. When traced to their source, these dubious offsets often encourage climate protection that would have happened regardless of the buying and selling of paper certificates. One danger of largely symbolic deals is that they may divert attention and resources from more expensive and effective measures.'

- The following week, on 3 April, London's *Independent* newspaper revealed that 'Europe's big polluters pumped more climate-changing gases into the atmosphere in 2006 than during the previous year, according to figures that show the EU's carbon trading system failing to deliver curbs'.
- Later in April 2007, after an exhaustive series on problems associated with carbon trading and offsets – including the headline 'Industry caught in carbon "smokescreen"' – *The Financial Times* editorialised in favour of taxes, against trading:

> While short-term politics favour markets, taxes would be better in the long term, because industry needs certainty for investments years hence. A government committing to painful taxes signals the seriousness of its intentions. Carbon taxes, offset by cuts in other taxes, are more difficult to eliminate than artificial markets. Carbon markets have other problems. Above all, they fix the amount of carbon abated, not its price. Getting the amount of emissions a little bit wrong in any year would hardly upset the global climate. But excessive volatility or unduly high prices of quotas on carbon emissions might disrupt the economy severely. Taxes create needed certainty about prices, while markets in emission quotas create unnecessary certainty about the short-term quantity of emissions.

- *The Guardian* newspaper headlined its June 2007 investigation of carbon trading with equal scorn: 'Truth about Kyoto: Huge profits, little carbon saved . . . Abuse and incompetence in fight against global warming . . . The inconvenient truth about the carbon offset industry'.
- The October 2007 issue of *Hedge Funds Review* carried a special report on carbon trading:

> One year ago conventional wisdom held carbon prices would rise steadily and indefatigably, yet since April 2006 the market has been short . . . Besides uncertainty, the main issue that is preventing more hedge funds from entering the market is the lack of liquidity. [According to] Werner Betzenbichler, head of carbon management services at TUV SUD Industrie Service, 'It is not possible to verify data accurately and equipment is outdated and not suited to current standards of measurement in many cases'.

At the same time, a stark warning came from Yvo de Boer, head of the United Nations' Intergovernmental Panel on Climate Change (IPCC), who 'does not rule out the possibility that the market could collapse altogether'.

Towards the end of 2006, the British government released *The Stern Review: The Economics of Climate Change*, which estimated that climate change will cost 5–20 per cent of global output at current warming rates. Nicholas Stern called for demand-reduction of emissions-intensive products, energy efficiency, avoiding deforestation and new low-carbon technology and insisted that carbon trading has a key role (Stern 2006).

At around the same time in Nairobi, the twelfth UN climate change COP included further endorsements of carbon trading, including by NGOs such as Oxfam. A new adaptation fund was established, but with resources reliant upon CDM revenues. Activists from the Gaia Foundation, Global Forest Coalition, Global Justice Ecology Project, Large Scale Biofuels Action Group, the STOP GE Trees Campaign and World Rainforest Movement condemned the COP's move to biofuels and genetically engineered timber technology, which are also promoted through the CDM.

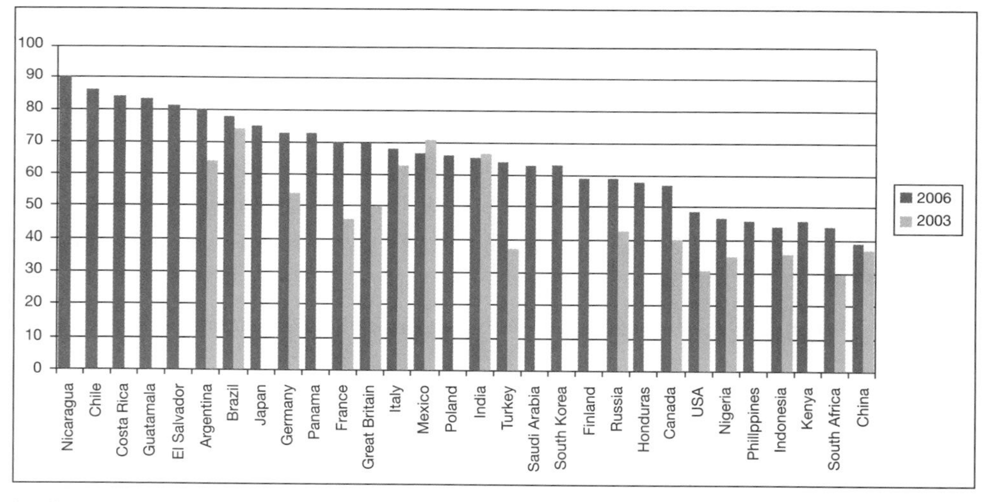

Is climate change a serious problem?

Source: GlobeScan, 2006

Climate change is a 'very serious' problem

No, for most South Africans it is not. The graph above is a measure of how many people in 30 countries believe climate change is a 'very serious' problem, according to surveyors

GlobeScan (2006). More than in nearly any other society, ordinary South Africans have been kept in the dark by the government, the media and business – with civil society making uneven efforts to address the deficit.

This level of ignorance is no accident, for the main argument we make in the pages that follow is that South African elites have adopted an exceptionally irresponsible posture by reinforcing the inherited apartheid economy's addiction to fossil fuels and indeed relying upon further cheap – yet for many, intermittently available – coal-fired electricity for the capital-intensive, extractive operations of mining firms, smelters and other industries, which mainly feed profits to multinational corporations (whose financial headquarters, some of which were once in Johannesburg, are now mostly abroad). This is the first problem.

The second problem is that the policy of giveaway electricity deals to crony capital has reduced Eskom's spare generation capacity to dangerously low levels and has compelled the government to move further down a very dangerous road to nuclear plant construction.

The third problem is that electricity is becoming unaffordable to poor and working-class people, with 50 per cent price hikes from mid-2000 levels by 2010, at precisely the moment it is most important to replace dirty household energy so as to assure gender equity, improve local economies, assist children in their studies and reduce respiratory illnesses (which take many people from being HIV-positive to full-blown AIDS).

The fourth argument, to top it all off, is that the South African government's positioning in global climate debates – such as Kyoto Protocol negotiations in Nairobi in 2006, Bali in 2007 and Accra in 2008 and obstructive alliances with the United States and Australia in-between – is self-interested and globally destructive, in so far as it adopts the premises we call 'the privatisation of the air' to address greenhouse gas emissions.

In tackling these problems, this book addresses climate change, the dangerous mitigation strategy most commonly termed carbon trading, as well as civil society reactions, including the crucial new countervailing strategy emerging in various sites: leave the oil in the soil, keep the non-renewable resources in the ground. We focus on the South African context we know best, but also aim to contribute to the global debate about how to address this crisis.

With climate change posing perhaps the gravest threat to humanity in coming decades and with free market economics still the global ruling elite's most powerful ethos, it is little wonder so much effort has gone into making the latter a solution to the former, no matter how much evidence has recently emerged that this is a strategy doomed to fail.

We believe South Africa is an extremely important site of study and action, given that our CO_2 emissions rate in the all-important energy sector – measured per person per unit of output (the economy's per capita energy intensity) – is twenty times worse than that of even the United States. It is also no wonder the South African government is such an enthusiastic promoter of carbon trading, given the systematic crony capitalism that characterises the corruption-ridden ruling party. We document this point later in the Introduction.

It is hard to understand how so many otherwise thoughtful journalists and commentators – even Yvo de Boer, head of the United Nations' IPCC – remain uncritical of the South African government's stance. The praise has flowed thick, especially after the Bali COP to the Kyoto Protocol in December 2007. And yet the conference was a disaster for the climate, as no firm targets were set and another long and fruitless round of negotiating with the United States now ensues. Hence the past fifteen years of climate talks since the Rio de Janeiro Earth Summit represent the equivalent of fiddling while Rome burns. And Bali once again disguised a more profound preliminary clash in the coming climate wars: not between Washington, DC and the rest of the world, but between those tinkering with mitigation via carbon trading – so fossil-fuel addiction can continue untreated – and those aiming to genuinely halt the supply of and demand for non-renewable fossil fuels.

A clear example of the former is the Kuyasa energy-efficient house retrofitting scheme in Khayelitsha township, Cape Town. By mid-2008, the project had begun to finally realise some of its potential (after a financing scandal brought to light in the US media). But Kuyasa's laudable, yet modest, renewable energy contributions at the micro-scale contrast with the extreme eco-hype it received from the world's eight most powerful politicians in 2005, who used Kuyasa to 'offset' emissions from their annual party, this time in Scotland. On 27 January 2007, *The Independent* newspaper reported the Blair government's failure to fund the G8's allegedly carbon neutral summit in Gleneagles (in July 2005) through a grant to Kuyasa, the first UN gold standard project:

> Two years on, Britain's £100 000 remains in the Treasury while the Kuyasa project struggles to get off the ground. The merits of carbon offsetting are increasingly being questioned by environmental experts. Critics argue governments, companies, even individuals, can pay for someone else to reduce their carbon emissions while doing nothing to cut their own carbon footprint. But as the problems faced by the Kuyasa project have now proved, it is not as straightforward as it may appear.
>
> The British Government has frequently highlighted the 'carbon neutral' G8 summit as an example of its commitment to tackling climate change. But the truth is very different. (Bloomfield 2007)

South Africa's contribution to climate change

South Africa is not included in the Kyoto Protocol Annex 1 list of countries that have to make emissions reductions and hence the economy as a whole is not subject to targets at this stage. But we will be in future and looking ahead, officials and corporations – and uncritical NGOs – are promoting the Kyoto Protocol's CDM as a way to continue South Africa's hedonistic output of greenhouse gases, while earning profits in the process. As shown in Appendix 1 in this volume, the Department issued the 'National Climate Change Response Strategy' in September 2004, insisting that society must understand 'up-front' how the 'CDM primarily presents a range of commercial opportunities, both big and small. This could be a very important source of foreign direct investment'.

But do we deserve to earn such 'investment' because of South African industry's indefensible contribution to global warming? From his base at the University of Zululand, Professor Mark Jury has gathered the following damning facts about South Africa's debt to the planet:

- South Africa contributes 1.8 per cent of total greenhouse gases, making it one of the top contributing countries in the world;
- the energy sector is responsible for 87 per cent of CO_2, 96 per cent of sulphur dioxide and 94 per cent of nitrous oxide emissions;
- 90 per cent of electricity is generated from the combustion of coal that contains more than 1 per cent sulphur and more than 30 per cent ash;
- with a domestic economy powered by coal, South Africa has experienced a five-fold increase in CO_2 emissions since 1950;
- South Africa is signatory to the UNFCCC and Montreal Protocol, yet CO_2 emissions increased 18 per cent between 1990 and 2000;
- South Africa has only recently enacted legally binding air pollution regulations via the National Environmental Management Air Quality Act, but energy efficiency is low;
- in rural areas of South Africa, approximately three million households burn fuel wood for their energy needs, causing deforestation, reduction of CO_2 sinks and health problems;
- the industrial sector consumes 2.6 Quads of energy (57 per cent of total primary energy consumption) and emits 66.8 megatonnes of carbon (65 per cent of total carbon emissions from fossil fuels), although industry's contribution to gross domestic product (GDP) is 29 per cent;
- since 1970, South Africa consistently has consumed the most energy and emitted the most carbon per dollar of GDP among major countries. South African energy intensity measured 33.5 Kilo British Thermal Units (KBTU) per US$ unit – nearly the same as China;

- South Africa's carbon intensity is far higher than in most other countries due to its dependence on coal; and
- household and industrial energy consumption across the continent is predicted to increase by over 300 per cent in the next 50 years, with significant growth in sulphur and nitrogen emissions (Jury 2004).

Coal is by far the biggest single South African contributor to global warming, representing between 80 and 95 per cent of CO_2 emissions since the 1950s. But liquid CO_2 emissions mainly from transport have risen to the level of more than 10 000 metric tonnes a year since the early 1990s. It is regrettable but true, just as in Eastern Europe (whose CO_2 emissions are well below 1990 levels) that the long recession of the early 1990s was the only point in South Africa's history since the economic crisis of the 1930s that CO_2 emissions stabilised and dropped slightly. Needless to say, South Africa is by far the primary global warming villain in Africa, responsible for 42 per cent of the continent's CO_2 emissions, more than Egypt, Nigeria, Algeria and Libya put together.

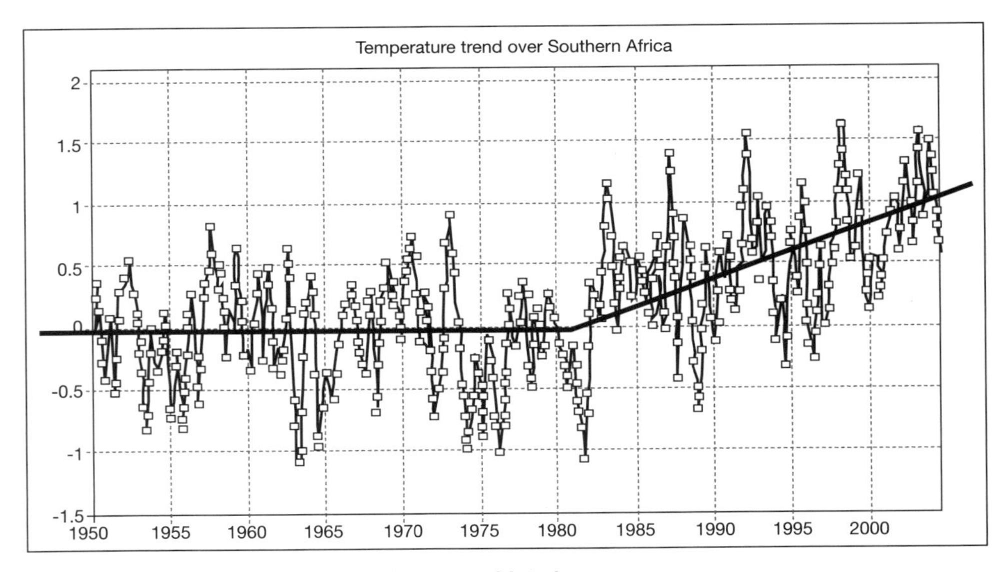

Rise/fall in southern African temperatures over historic norms.

Source: Mark Jury

Given the vast increase in CO_2 emissions by South Africa, especially during the 1980s–90s, added to similar increases in global greenhouse gas emissions, it is only logical to find an average of 1 degree Celsius increase in our region's temperature, over historic norms. This is merely the surface-level statistical information about the climate change crisis, as it emerges. Much more could be said about the various other indicators, ranging from droughts and floods in South Africa and Africa to the hurricanes which belted George W. Bush's oil-producing and refining belt in Texas and Louisiana in September 2005.

To critics of the Kyoto Protocol, including dozens of environmental justice networks that signed the October 2004 'Durban Declaration on Climate Justice', the CDM and especially the new carbon market that permits trade in pollution rights represent misleading 'greenwash'. Carbon trading justifies letting the United States, the European Union and Japan continue their emissions, in exchange for a small profit payout to unscrupulous South African firms and municipalities for reductions in local carbon – reductions that we should already be making.

For example, methane that escapes from Africa's largest landfill, at Bisasar Road in the Durban residential suburb of Clare Estate, should be captured, cleaned and safely turned into energy. eThekwini officials instead aim to burn the methane on site and in the process, keeping the toxic dump open at least another seven years – although the African National Congress (ANC) had promised its closure in 1996 due to community opposition. The officials' goal is to sell carbon credits via the World Bank to big corporations and Northern governments. Although a famous community activist, cancer-stricken Sajida Khan, frightened the World Bank off initially with a hard-hitting environmental impact assessment (EIR), she died of her illness in July 2007 and it appears that the eThekwini municipality has every intention of continuing the project with new partners.

The South African economy's five-fold increase in CO_2 emissions since 1950 (and 20 per cent increase during the 1990s) can largely be blamed upon the attempt by Eskom, the mining houses and metal smelters to boast of the world's cheapest electricity. Furthermore, there are very few jobs in these smelters, including the proposed US$2.5 billion Coega aluminium project, for which the notorious Canadian firm Alcan has been promised lucrative sweetheart deals from Eskom, the Department of Trade and Industry and the Industrial Development Corporation (IDC). Less than 1 000 jobs will be created in the smelter, although it will consume more electricity than the nearby city of Port Elizabeth (Mandela Metropole).

Aside from carbon trading, the main answer to the climate question provided by public enterprises minister Alec Erwin is fast-tracking the dangerous and outmoded pebble bed technology rejected by German nuclear producers some years ago – a reckless strategy that will continue to be fought against by Earthlife, the civil society group that won two important preliminary court battles against Erwin's special adviser, former director-general of the Department of Environmental Affairs and Tourism (DEAT), Chippy Olver.

Instead, renewable sources such as wind, solar, wave, tidal and biomass are the only logical way forward for this century's energy system, but still get only a pittance of government support, a fraction of the hundreds of millions of rand wasted in nuclear research and development. Meantime, because of alleged 'resource constraints', communities such as Kennedy Road bordering the Bisasar landfill – where impoverished people rely upon dump scavenging for income – are still denied basic services like electricity. While Kennedy Road activists are promised a few jobs and bursaries from the CDM proceeds, the plan to burn the landfill's methane gas onsite could release a cocktail of new toxins into the already poisoned air. Gas flaring would increase fifteen-fold under the scheme that eThekwini has tried selling to the World Bank. The generator's filters would never entirely contain the hydrocarbons, nitrous oxides, volatile organic compounds, dioxins and furans.

An even more dubious carbon trade is now being marketed: Sasol's attempt to claim credits for its new Mozambique gas pipeline, on the grounds that the huge investment would not have happened without it. That this is a blatant fib was conceded offhandedly to researchers by a leading Sasol official in June 2005, within four months of the Kyoto Protocol. The gas pipeline project proposal to the Protocol's CDM lacks the key requirement of additionality – the firm doing something (thanks to a lucrative incentive) that it would not have done anyway – thus unveiling the CDM as vulnerable to blatant scamming. This is not the first blatant spindoctoring or outright fraud associated with CDM claims in South Africa or elsewhere and it is the sort of incident that discredits the whole idea of commodifying the air through unverifiable carbon reductions.

Aside from the World Bank, the cash-rich companies that most need to cut these deals to protect their future rights to pollute are the oil majors, beneficiaries of windfall profits as the price per barrel soared from US$11 in 1998 to more than US$70 in 2006. The Bank itself even admits in a recent study that these and other extractive firms' depletion of Africa's natural resources drain the national wealth by hundreds of dollars per person each year (World Bank 2005).

In the process, the oil fields are attracting a new generation of US troops to bases being developed in the Gulf of Guinea. Once again, the South African government is amplifying the worst trends, as Human Sciences Research Council (HSRC) researchers John Daniel and Jessica Lutchman conclude of sleazy oil deals – not only by Imvume in Saddam Hussein's Iraq, replete with transfers to ruling party coffers, but also involving the Sudanese and Equatorial Guinean dictatorships: 'In its scramble to acquire a share of this market, the ANC government has abandoned any regard to those ethical and human rights principles which it once proclaimed would form the basis of its foreign policy' (Daniel and Lutchmann 2005). President Thabo Mbeki himself downplayed Sudan's Darfur crisis, even when sending peace-keeping troops, because, as he said after a meeting with Bush in mid-2005, 'If you denounce Sudan as genocidal, what next? Don't you have to arrest the president? The solution doesn't lie in making radical solutions – not for us in Africa' (Becker and Sanger 2005). The national oil company, PetroSA, had five months earlier signed a deal to share its technicians with Sudan's Sudapet, so as to conduct explorations in Block 14, where it enjoyed exclusive oil concession rights (Fabricius 2005).

The ethical principles referred to by Daniel and Lutchmann should be urgently revisited now, since our future generations' very survival is at stake. Since the Department of Environmental Affairs and Tourism's October 2005 National Climate Change Conference did not engage seriously with these critiques, its powerbrokers should be regarded as a large part of the problem. The irony is that while generating enormous carbon emissions, energy is utilised in an extremely irrational way. The unjust system leaves too many people without access to energy, while a few large corporations benefit disproportionately.

In sum, our purpose is to dig deeper, in order to uncover an emerging form of environmental injustice – the carbon market – and to highlight cutting-edge attempts to mitigate that injustice through civil society activism and advocacy that stretches from retail household reconnections to international environmental negotiations. This is not an entirely celebratory account, for notwithstanding successful civic resistance to South Africa's largest proposed project, at Bisasar Road, one of the concerns our research has uncovered is the failure of the environmental justice critique to penetrate the realm of policy. In that sphere, Big Oil and the South African 'minerals-energy complex' (Fine and Rustomjee 1996) appear to have the upper hand.

The role of the South African government

As environment minister for the South African government in the mid-2000s, Marthinus van Schalkwyk underwent an extreme makeover from the man known as *Kortbroek* (literally, 'short pants', a reference to his youth and boyish demeanour). A former student spy for the government during apartheid and then late 1990s National Party leader after F.W. de Klerk, Van Schalkwyk liquidated the once formidable racist political force into the ruling ANC, so as to earn himself (and no one else from the party) a cabinet post.

In September 2006, at a Western Cape meeting of the IPCC, Van Schalkwyk rated CDM promotion second in his three priorities for the upcoming Nairobi COP meeting (between more adaptation funding and tougher targets for Kyoto): 'The 17 CDM projects in the pipeline in sub-Saharan Africa account for only 1.7 per cent of the total of 990 projects worldwide. To build faith in the carbon market and to ensure that everyone shares in its benefits, we must address the obstacles that African countries face' (Van Schalkwyk 2006a: 2).

After Nairobi, the minister wrote in *Business Day* that South Africa had achieved its key objectives, including kick-starting the CDM in Africa and welcomed UN support for more 'equitable distribution of CDM projects', concluding that this work 'sends a clear signal to carbon markets of our common resolve to secure the future of the Kyoto regime' (Van Schalkwyk 2006b). Yet within days, on 24 November, Van Schalkwyk's colleagues confirmed the largest industrial subsidies in African history – entailing a vast increase in coal-fired electricity – for the Coega smelter and export processing zone near Port Elizabeth (Mandela Metropole). Assuming supplies are assured (which after load-shedding began in January 2008, became a crucial deterrent), Alcan plans to build a US$2.5 billion aluminium smelter, thanks to inexpensive electricity.

The following week, University of Cape Town's environmental studies professor, Richard Fuggle, attacked the CO_2 emissions associated with the Coega deal in his retirement speech, describing Van Schalkwyk as a 'political lightweight' who is 'unable to press for environmental considerations to take precedence of "development"' (Gosling 2006).

Just over three months later, in March 2007 Van Schalkwyk issued the South African government's 'climate change roadmap', based on the need to make 'mitigation policies and measures part of a pro-development and growth strategy' (2007a). Yet the essence of the roadmap was business as usual: 'Rather than viewing action on climate change as a burden, the message is that action on climate change also holds myriad opportunities for new investment in climate friendly technologies, creating access to cleaner energy for development and building new competitive advantages in clean and renewable technologies'. Moreover, in listing co-operating state agencies, he failed to even mention the two most responsible for the South African economy's world leadership in CO_2 emissions: the Department of Trade and Industry and the Treasury.

Reacting to the strategy in a *Mail & Guardian* interview, Nicholas Stern observed, usefully, 'South Africa's over-reliance on coal' requires 'innovative ways to reduce the heavy carbon footprint' and 'climate change would be an economic disaster for South Africa'. But, less convincingly, he suggested that carbon capture technology would be a solution and that Van Schalkwyk was 'an important figure in global discussions on climate change and South Africa had the potential to bring opposing factions such as China and the US together' (Groenewald 2007).

The hope that carbon capture and storage might become an even greater part of South Africa's arsenal was also advanced in the DEAT's long-term mitigation scenario exercise conducted in 2007, but released in mid-2008. The 'Use the Market' component of this scenario wistfully projected the following merits of 'an escalating CO_2 tax, or alternative market mechanism' (unspecified, but possibly an internal emissions market): 'Impact on

GDP is mildly positive (0.73%) instead of the previous minus 2%; price increases are overshadowed by higher investments; income from employment increases for all household groups; and differences in welfare effects are marginal' (DEAT 2008: 26–29). We may look back and find these as whimsical as 'Government's vision for the road ahead on climate change', which is that 'emissions must peak, plateau and decline – stop growing at the latest by 2020–2025, stabilise for up to ten years, then decline in absolute terms'. This is possible, DEAT argues, if South Africa can 'redefine our competitive advantage and structurally transform the economy by shifting from an energy-intensive to a climate-friendly path as part of a pro-growth, pro-development and pro-jobs strategy'. And yet as those words were written, the public enterprises minister, Alec Erwin, was announcing R1 trillion in new coal and nuclear energy investments. The only indication that Van Schalkwyk was paying attention was the claim (with no details) that the new coal-fired power plants should include 'carbon capture readiness' prior to approval.

This midwife role between major polluters – in which Van Schalkwyk sought a deal that would not endanger status quo interests (especially the carbon trading and nuclear industries) – culminated in Bali, where a non-binding treaty was agreed that lacks even elemental specifications about what needs to be done to halt global warming. Van Schalkwyk's own role as climate negotiator for the G77 bloc of Third World countries was set to continue after Bali, towards the 2009 Copenhagen summit, which is meant to replace Kyoto with a 2012–20 emissions reduction strategy. But whether he survives the vast fallout expected from Mbeki's defeat in the ANC leadership race in December 2007 remains to be seen. However, Jacob Zuma promised Citibank and Merrill Lynch in November 2007 that we can expect no basic policy change when and if (for corruption charges haunt his ANC leadership) he takes state power (Bond 2007). Hence South Africa will most likely continue to trade carbon and promote nuclear energy, while ignoring the contradictions associated with its own worsening emissions. However, one of the variables that could become relevant in the government's future policy is resistance to orthodox climate management, both within and outside South Africa.

Van Schalkwyk is an unlikely leader of the Third World against the North, although that was the reputation he gained in Bali. For standing up to the belligerent US negotiator, editorialists at the *Sunday Independent* (23 December 2007) favourably cited an unnamed observer's view: 'One could perhaps say he has turned from *Kortbroek* to *Langbroek*.'

More typically, a few weeks earlier, Van Schalkwyk went to Washington, DC for the International Emissions Trading Association (IETA) forum, arguing in favour of carbon trading on grounds that 'an all-encompassing global carbon market regime which includes all developed countries is the first and ultimate aim'. Without apparent irony, he appealed to 'individual nations to rise above short term self-interest for the benefit of the long term common good' (2007b: 1).

Within days, the parastatal Central Energy Fund (CEF) announced a new carbon trading fund based in London, CEF Carbon Markets. Its director, Deven Pillay, told audiences about 'billions' in carbon trading deals that will materialise in South Africa, even though there are only nineteen deals underway, which include some lemons described in Part 2 of this book.

One journalist's account (in the *Mail & Guardian*) of this development made a telling point but then lost the plot: 'the CEF of old did little more than house the country's Strategic Fuel Fund and was notorious for paying inflated premiums to middlemen, such as Marino Chiavelli, Marc Rich and John Duess' (Davie 2007).

It is because of this approach, akin to pimping South Africa's exceptionally high carbon emissions for self-interested profits at the expense of the local and global climate, that we advocate a far more sceptical view of the South African government and the emerging market-based carbon management regime. After all, the Washington Consensus did enormous damage to South Africa's economy, beginning when the National Party adopted International Monetary Fund (IMF) advice in the late 1980s, advice enthusiastically adopted in 1993 in the government of national unity's first action – an IMF structural adjustment programme by stealth – and then amplified when the ANC took power in 1994. The vast costs and current economic vulnerabilities are readily demonstrated, as shown at the end of Chapter 2.

Somehow, however, the mainstream viewpoint regarding both macroeconomic management and climate change lauds South Africa's Washington, DC-friendly managers Thabo Mbeki, Trevor Manuel, Alec Erwin and Marthinus van Schalkwyk, even while mass grass-roots protests rose to new heights in the mid-2000s, up to 30 on an average day, according to the South African Police Service.

Regarding climate policy, the same uncritical perspective prevails. On the one hand, Durban's *The Mercury* newspaper editorialised on 19 December, a few days after Bali ended: 'South Africa has set an example by serving notice in Bali that, as a developing country, we too are prepared to shoulder our share of the responsibility for taking on this global threat'. On the other hand, editorialists at its sister paper *The Sunday Independent* acknowledged four days later:

> However well intended Van Schalkwyk's undertakings before the 180 countries represented at the UN Framework Convention on Climate Change in Bali might have been, fact is that, like China and India, South Africa is embarked on a radical programme to increase its electricity output. The generating source will be coal, which, with Sasol's mammoth oil-from-coal process added, is one of our biggest producers of greenhouse gas.

Van Schalkwyk's failure to question the broader process means the South African government will become much more reliant upon the two most reactionary climate mitigation strategies: carbon trading and increased nuclear energy (Chapter 3 tackles the nuclear energy threat).

Resistance in South Africa

The first critics of climate trading were not from big business and its journalists, which stand to lose substantial sums as the various scam reduction projects are exposed. Demonstrations were already being held by Rising Tide and other activist groups against carbon meetings in the early 2000s and in 2003, the Amsterdam-based Transnational Institute's Carbon Trade Watch issued *The Sky is Not the Limit* – a pamphlet featuring Sajida Khan's Bisasar Road struggle – to warn of the impending problems. By October 2004, the 'Durban Declaration on Carbon Trading' (Appendix 2 in this volume) was produced not far from the home of Khan, who provided dozens of the core signing group with an inspiring example of local resistance to the new carbon market.

What became known as the Durban Group for Climate Justice subsequently established one of the finest expert-knowledge networks in support of grass-roots struggles across the Third World, with powerful voices from Indonesia, Thailand, India, South Africa, Brazil and Ecuador providing political guidance and unveiling CDM damage to allied researchers and campaigners in think tanks and advocacy groups including The Corner House, Carbon Trade Watch, the Institute for Policy Studies' Sustainable Energy and Economy Network (SEEN), SinksWatch, Dartmouth College's Environment Department and the University of KwaZulu-Natal's Centre for Civil Society (CCS).

The Durban Group for Climate Justice's original sponsor, the Dag Hammarskjöld Foundation of Uppsala, Sweden, published Larry Lohmann's monumental book *Carbon Trading* in late 2006 and within nine months had recorded more than 333 000 downloads and distributed over 10 000 hard copies. Booklets were prepared by the Transnational Institute's Carbon Trade Watch, with titles such as 'Where the Trees are a Desert: Stories from the Ground' (2003), 'Hoodwinked in the Hothouse: The G8, Climate Change and Free-Market Environmentalism' (2005), 'Agrofuels – Towards a Reality Check in Nine Key Areas' (2007a) and 'The Carbon Neutral Myth: Offset Indulgences for your Climate Sins' (2007b). Other books about South Africa were produced (including this one) and three specialist videos were made about Sajida Khan's struggle (all available on the DVD set *CCS Wired*), including one by Rehana Dada, which aired in 2006 on the South African Broadcasting Corporation's (SABC's) environmental show *50/50*.

These are invaluable resources in our collective efforts to ensure genuine cuts in emissions, including leaving oil in the soil and other non-renewable resources in the ground. They help us to understand one of the final frontiers of the commodification of everything: the air itself.

The structure of this book

We begin this volume with core argumentation and context, in Part 1. Patrick Bond opens with a critique of the South African national energy system and Muna Lakhani and Vanessa Black add to this in Chapter 2 with more detail about the government's dangerous nuclear fantasy and its underfunding of vitally needed renewable energy sources.

In Part 2, South Africa's recent experience with carbon trading is analysed by Graham Erion, with the assistance of Larry Lohmann and Trusha Reddy.

In Part 3 we ask: who benefits from the status quo? We are grateful for assistance from Durban Group for Climate Justice colleagues at Amsterdam's Transnational Institute Carbon Trade Watch and the Institute for Policy Studies SEEN. In Chapter 4 Heidi Bachram articulates concerns over carbon colonialism, while in Chapter 5, Daphne Wysham and Janet Redman provide a critique of World Bank strategies. Others associated with the 'Durban Declaration on Carbon Trading' – Lohmann, Kill, Erion and Dorsey – unveil the PCF beneficiaries. The main beneficiary of carbon trading is Big Oil. In September 2005, an exceptionally powerful analysis was presented to the World Petroleum Congress by the Pietermaritzburg-based NGO groundWork, published as 'Whose Energy Future? Big Oil against the People of Africa'. We are grateful for this world-renowned NGO's research and permission to excerpt from that book (see Chapter 7). At the end of Part 3, Bond shows how the search for Africa's oil has generated serious geopolitical and economic crises for the continent's citizens.

In Part 4 we consider resistance, including two analyses of the global situation written at the end of 2007: from Joan Martinez-Alier and Leah Temper on the broader debate about climate options (an article which was originally published in India's *Political and Economic Weekly*) and from Brian Tokar on civil society strategies beyond Bali and beyond mere condemnation of carbon trading. As Bond concludes in Chapter 11, after revisiting the debate in mid-2008, one crucial strategy is that Africans and all people who live above or near fossil-fuel deposits, should consider leaving the oil in the soil and retaining other non-renewable resources ranging from minerals to old-growth forests intact, since their exploitation so often *under*develops countries and also so clearly threatens the world's climate. To achieve this logical outcome will require a far stronger international push to limit Big Oil's power, not to mention the Bush imperial agenda. It will require us all to work overtime for reparations – US$75 billion per year merely for serving as a carbon sink, experts argue – that the South is owed by the North. This project continues, not only within the global justice movements, but also in other CCS work still to be published.

Our audience

In gathering material for this book, our central priority was to generate debate with what might be considered the 'reform' wing of the climate activist community who, through networks like the Climate Action Network (CAN) and the South African Climate Action Network (SACAN), have since 1997 accepted carbon trading as a necessary evil.[1] One ethnographer of the environmental NGO industry, Michael Dorsey, wrote from the Bali negotiations:

> The largest NGO here is the International Emissions Trading Association – IETA. With 336 representatives, they account for just under 8 percent of all the NGO delegates, more than the delegations of Greenpeace, the World Wildlife Fund and Friends of the Earth combined! What is particularly disturbing is that after more than a decade of failed politicking, many NGO types have not only failed to make progress, but in the midst of the failures are only partially-jumping off the sinking ship – so as to work for industries driving the problem. Unfortunately, many continue to influence NGO policy, from their current positions, while failing to admit to or even understand obvious conflicts of interest. (personal communication, 10 December 2006)

Fortunately, glaring NGO co-optation is not as blatant in South Africa. However, at the highest levels of environmental policy, former minister Vali Moosa has earned a dubious reputation at the International Union for the Conservation of Nature (IUCN), over which he presides, not only for chairing Eskom's board at an inopportune time as it develops nuclear technology and fails to roll out electricity to the rural poor, but also because he has an active carbon trading business.

There is some hope that South African NGOs can rise to the challenge. In October 2005 at the National Climate Change Conference in Johannesburg (in the wake of the publication of our book *Trouble in the Air*, which showed the dangers of the South African carbon trade), SACAN did at least resolve to:

> Raise awareness of the limitations of the Clean Development Mechanism, the undue emphasis placed on carbon trading in strategies for renewable energy and efficiency and the need for rigorous application of the sustainable development criteria by the designated national authority to ensure that negative local impacts are not offset against global benefits. (SACAN National Climate Change Resolution 2005)

However, years have since passed, without SACAN doing the promised consciousness-raising, hence the ongoing ability of Van Schalkwyk to push carbon trading as if it were a genuine solution.

Whether there are genuine 'global benefits' remains to be seen, for at best, CDMs represent a shifting of the deck chairs on the climate Titanic. A better example for SACAN would be the struggle over the CDM proposal at Durban's Bisasar Road, where more critical climate justice activists halted the potential US$15 million carbon trading project.

To be fair, however, it is also true that reform-minded environmental non-governmental organisations (ENGOs) – including leadership at SACAN – have been far more effective in pointing out the problems in the strategy, even as they seek to improve it. Nevertheless, there are far too many staff and members of the large, corporate-funded international NGOs such as the IUCN, Sierra, the World Wildlife Federation, the Environmental Defence Fund and even Greenpeace who have bought into carbon trading – although exceptional individuals do struggle against the current of market environmentalism.

Consider a fairly typical comment from the market-friendly camp (Sierra Club Canada director Elizabeth May): 'I would have preferred a carbon tax, but that is not the agreement we have. The reality is that Kyoto is the only legally binding agreement to reduce greenhouse gases. When you're drowning and someone throws you a lifeboat, you can't wait for another one to come along' (Athanasiou 2005). Our rebuttal is that carbon trading is simply not a seaworthy lifeboat and as temperatures (and sea levels) rise, we are discovering, to our peril, the numerous leaks. It is largely for the sake of these market environmentalists and for their constituents who deserve wider advocacy choices that we have gathered the damning information about the carbon trade described in the pages that follow.

To illustrate the importance of education (of the type done by our colleagues in Carbon Trade Watch), here is a paragraph written by Andrew Leonard, whose 'How the World Works' column is familiar to readers of the popular *Salon* e-zine:

> The activists behind Carbon Trade Watch are smart, and their critique of the carbon trading system is brutal and effective. How the World Works has long been attracted to market mechanisms that would create financial incentives for reducing pollution, but after pondering the arguments marshalled by Carbon Trade Watch, we feel our optimism melting away like Greenland's glaciers.
>
> Some environmental advocacy defeats its own purposes by transparently manipulating facts to fit its own agenda, a strategy that raises hackles even when you agree with the overall cause. But Carbon Trade Watch, although clearly in pursuit of an agenda, doesn't make that mistake. In particular, the

> connections drawn between the root causes of climate change and the forces propelling globalisation are compelling – for example, a vast proportion of the investment capital pouring into middle-income emerging nations like Brazil, India, China and South Africa is going directly to non-renewable energy development, with consequences for global warming that dwarf any reductions in emissions the Kyoto Protocol may accomplish, even if it works as planned. (Leonard 2006)

The implications of our findings are also clear for broader South African society and also for those government officials who will probably continue to act as irresponsibly as US officials, if the National Climate Change conference in Johannesburg in October 2005, the Kyoto Protocol negotiations in Montreal in December 2005 and Nairobi in November 2006 are anything to go by.

All South Africans must face up to their responsibility for permitting the country's ruthless powerbrokers – mining/smelting magnates such as the late Brett Kebble, the Oppenheimer family, Lakshmi Mittal and newly enriched Patrice Matsepe, Tokyo Sexwale and Mzi Khumalo – to befriend the ANC government. The result is that the state provides them electricity prices at the world's lowest levels, largely for the benefit of mining or smelting empires, whose profits and dividends now mainly flow from South Africa to Britain, the United States and Australia. The recent upsurge in earnings by mining and smelting firms is a good indicator of the extent to which South African public policy and greed are threatening our descendants' very lives. However, we believe that this policy can be reversed, if even a little foresight is exercised and if political will is marshalled to contest the present adverse balance of forces.

Note

1. Unfortunately, some CAN leaders developed personal conflicts of interest which help explain why, in the face of so much evidence, they persevere with carbon trading promotion. This list was compiled by Michael Dorsey:
   - One board member, Jennifer Morgan of the Worldwide Fund for Nature, took leave for two years to direct work on Climate and Energy Security at a private firm, E3G, active in promoting the trade.
   - Another CAN activist, Kate Hampton of Friends of the Earth, joined a trading group, Climate Change Capital, as head of policy, as well as advisor to the EU High Level Group on Competitiveness, Energy and Environment and the California Economic Partnership Agreement (EPA) Market Advisory Committee.
   - Steve Sawyer was formerly the leader of Greenpeace in the United States and internationally, but moved over to lead the Global Wind Energy Council, a strong CDM promotion agency.

- John Sohn was formerly at Friends of the Earth and the World Resources Institute, but left for Climate Change Capital.
- Bryony Worthington was formerly at Friends of the Earth and then moved to the job of sustainable development manager at Scottish & Southern Power Corporation.
- Tracy Jones of the Union of Concerned Scientists simultaneously consults on CDM projects including timber.
- Catherine Pearce was formerly international climate campaigner at Friends of the Earth, but moved to the Parliamentary Renewable and Sustainable Energy Group, which serves UK politicians and senior industry stakeholders, including ConocoPhillips European Power and Shell International.
- James Cameron was a founding director of the Center for International Environmental Law and then went on to found Climate Change Capital and head its policy unit and advisory board.

PART 1

SOUTH AFRICA'S ENERGY CRISES

1

Dirty Politics

South African Energy

Patrick Bond[1]

There is perhaps no better way to interpret power relations in contemporary South Africa than by examining who has had access to energy in the past, who is getting it now and at what cost and who will have it in the future. The argument below is that the larger players in the energy market – i.e., transnational capital, accommodating neoliberal multilateral agencies and national governments and the rich – are having a disproportionate effect on public policy, even in South Africa.

Contradictions abound, of course. For Anton Eberhard of the National Electricity Regulator, there is 'no simple transition from a state centred electricity supply industry to an idealised World Bank electricity supply industry model' (Eberhard n.d.). Of course, the 'idealised World Bank model' has failed nearly everywhere, not only in electricity and energy (and especially electricity), but across the board.

Hence it is no surprise that during the transition to energy neoliberalism, the core components of South Africa's energy system are beset by anti-social, anti-ecological practices. These include climate change caused by what Fine and Rustomjee have termed the 'minerals-energy complex' (1996), the crisis of electricity access in view of disconnections associated with energy sector liberalisation and the government's failure to promote renewable energy sources, while wasting scarce funds on a nuclear energy fantasy. However, the most important issues to flag at the outset are the commodification of all life and nature implicit in carbon trading and the extraordinarily cheap supply of electricity that South African corporate users enjoy.

Commodification through carbon trade

The most clear – and simultaneously terrifying – commitment to the commodification of nature came from Lawrence H. Summers, then the chief economist of the World Bank, later US Treasury secretary and then president of Harvard University. There is merit in a full reprint of his World Bank memo (leaked to *The Economist* and published – with an endorsement – on 8 February 1992), merely to indicate the reasoning that helped to justify emissions trading at the time of its origins, in the run-up to the June 2002 United Nations conference on sustainable development (the Rio de Janeiro Earth Summit). At that conference, the World Bank took global intellectual and financial leadership on many environmental fronts and hence the consequences in Durban follow the eco-racist strategy Summers lays out here:

> DATE: December 12, 1991
> TO: Distribution
> 'Dirty' Industries: Just between you and me, shouldn't the World Bank be encouraging MORE migration of the dirty industries to the Lesser Developed Countries (LDCs)? I can think of three reasons:
>
> 1) The measurements of the costs of health impairing pollution depends on the foregone earnings from increased morbidity and mortality. From this point of view a given amount of health impairing pollution should be done in the country with the lowest cost, which will be the country with the lowest wages. I think the economic logic behind dumping a load of toxic waste in the lowest wage country is impeccable and we should face up to that.
> 2) The costs of pollution are likely to be non-linear as the initial increments of pollution probably have very low cost. I've always thought that under-populated countries in Africa are vastly UNDER-polluted, their air quality is probably vastly inefficiently low compared to Los Angeles or Mexico City. Only the lamentable facts that so much pollution is generated by non-tradable industries (transport, electrical generation) and that the unit transport costs of solid waste are so high prevent world welfare enhancing trade in air pollution and waste.
> 3) The demand for a clean environment for aesthetic and health reasons is likely to have very high income elasticity. The concern over an agent that causes a one in a million change in the odds of prostrate cancer is obviously going to be much higher in a country where people survive to get prostrate cancer than in a country where under 5 mortality is 200 per thousand. Also, much of the concern over industrial atmosphere discharge is about visibility impairing particulates. These discharges may have very little direct health impact. Clearly trade in goods that embody aesthetic pollution

> concerns could be welfare enhancing. While production is mobile the consumption of pretty air is a non-tradable.
>
> The problem with the arguments against all of these proposals for more pollution in LDC's (intrinsic rights to certain goods, moral reasons, social concerns, lack of adequate markets, etc.) could be turned around and used more or less effectively against every Bank proposal for liberalisation.

Summers aims to fix a market-caused problem with a market solution, no matter how unjust it is and regardless of the threat to human and environmental existence represented by the commodification of everything that he advocates. Although Summers did not get the position of chair of the council of economic advisers in the Clinton administration because of the memo above (Al Gore vetoed the position), he did rise to much more prominence in the late 1990s as US Treasury secretary. In 1998, a new ban on hazardous waste trade was promoted by the United Nations (UN) via a strengthened Basel Convention, but along with governments in Canada, Australia and New Zealand, Clinton refused to sign it. When the November 2000 presidential elections in Florida demonstrated banana republic tendencies in the United States, Summers was forced to take up the presidency of Harvard University.

Although he suffered the humiliation of losing his Harvard presidency in early 2006 as a faculty vote of no confidence loomed, Summers might look proudly upon an incident in mid-2006, showing the durability of his advice for low-wage countries:

> One August morning, people living near the Akouedo rubbish dump in Abidjan, capital of the Ivory Coast, woke up to a foul-smelling air. Soon, they began to vomit, children got diarrhoea, and the elderly found it difficult to breathe. 'The smell was unbelievable, a cross between rotten eggs and blocked drains', said one Abidjan resident. 'After 10 minutes in the thick of it, I felt sick.'
>
> As they live near the biggest landfill in Abidjan, the people of Akouedo are used to having rubbish dumped on their doorstep. Trucks unload broken glass, rotting food and used syringes. Children try to make the best of their dismal playground, looking for scraps of metal and old clothes to sell for a few cents. But this time, the waste would benefit no one. By yesterday, at least six people, including two children, had died from the fumes. Another 15 000 have sought treatment for nausea, vomiting and headaches, queuing for hours at hastily set up clinics. Pharmacies have run out of medicines and the World Health Organisation has sent emergency supplies to help the health system. The Ivorian government had resigned over the matter and, so far,

> eight people have been arrested. The tragedy is said to have begun on 19 August, after a ship chartered by a Dutch company offloaded 400 tons of gasoline, water and caustic washings used to clean oil drums. The cargo was dumped at Akouedo and at least 10 other sites around the city, including in a channel leading to a lake, roadsides and open grounds . . .
>
> Probo Koala, the ship that offloaded the waste, is registered in Panama and chartered by the Dutch trading company Trafigura Beheer. Trafigura had tried to offload its slops in Amsterdam, but the Amsterdam Port Services recognised its contents as toxic and asked to renegotiate terms. Trafigura said shipping delays would mean penalties of at least $250 000 so handed it over to a disposal company in Abidjan alongside a 'written request that the material should be safely disposed of, according to country laws, and with all the correct documentation'.
>
> This story is a common one. All down the West Africa coast, ships registered in America and Europe unload containers filled with old computers, slops, and used medical equipment. Scrap merchants, corrupt politicians and underpaid civil servants take charge of this rubbish and, for a few dollars, will dump them off coastlines and on landfill sites. (Selva 2006)[2]

Durable minerals/energy dependency

In addition to the commodification of nature and the use of Africa as a sink for Northern pollution, as indicated above, a second overarching problem addressed in this book is the centrality of cheap electricity to South Africa's economy, which stems from the power needs of mines and heavy industry, especially in beneficiating metallic and mineral products through smelting. *The Political Economy of South Africa* by Ben Fine and Zav Rustomjee puts the parastatal (Eskom) into economic perspective (1996). Here we locate electricity at the heart of the economy's minerals-energy complex, a 'system of accumulation' unique to this country. Throughout the twentieth century, mining, petro-chemicals, metals and related activities, which historically accounted for around a quarter of gross domestic product (GDP), typically consumed 40 per cent of all electricity, at the world's cheapest rates.

South Africa's largest parastatal firm, Eskom, plays a triple role, as: (a) generator of virtually all of the country's electricity, (b) sole transmitter and (c) distributor to many large corporations, municipalities, commercial farms, and to half South Africa's households, from sections of the largest municipalities to most rural villages. Eskom was crucial to South Africa's rapid capital accumulation during the past century. At the same time, as Fine and Rustomjee show, the company fostered a debilitating dependence on the (declining) mining industry. Economists refer to this as a 'Dutch disease', in memory of the damage done to

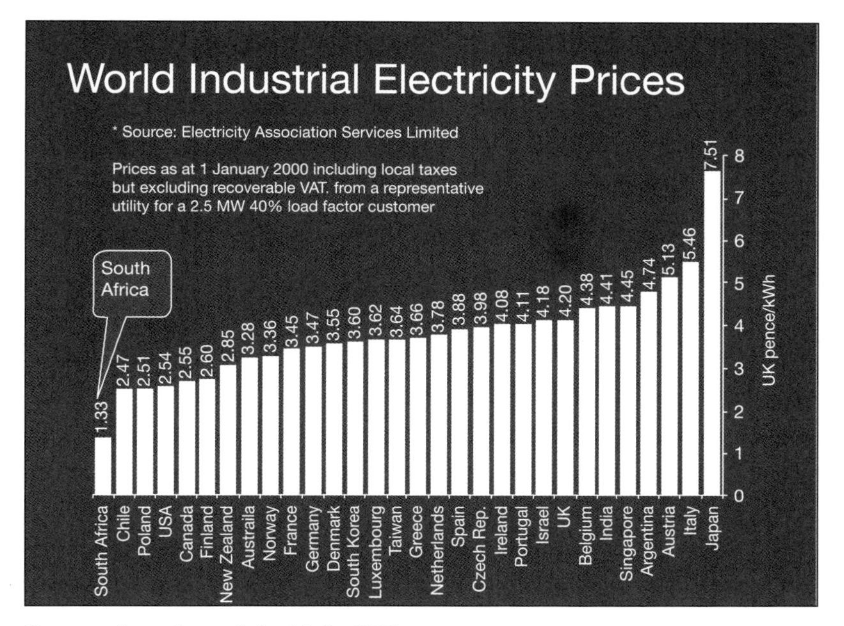

Comparative prices of electricity, 2000.

Source: Anton Eberhard, n.d.

Holland's economic balance by its cheap North Sea oil. Moreover, Eskom as the monopoly electricity supplier played a role in strengthening private mining capital by purchasing low-grade coal from mines that were tied to particular power stations on the basis of a guaranteed profit.

After the Second World War, growing demand from new mines and manufacturing caused supply shortages and resulted in a programme for the construction of new power stations. In the process, the apartheid state promoted Afrikaner-owned coal mines, with Eskom contracting these for a portion of its coal supply. The national grid – which linked previously fragmented power station supplies via transmission lines – was initially formed in 1964 and extended supply into the southern African region. Until 1985, when sanctions made international borrowing more difficult, foreign loans were used to build Eskom's massive excess capacity through environmentally damaging coal-fired power stations. At its peak in 1990, Eskom produced three-quarters of the African continent's electricity and its capacity was being extended to more than 37 000 megawatts at a time when peak demand was less than 25 000 megawatts (Clarke 1991: 33).

Eskom's power plants continued providing artificially cheap electricity to large, energy-intensive corporations and white households, including a new wave of subsidised white

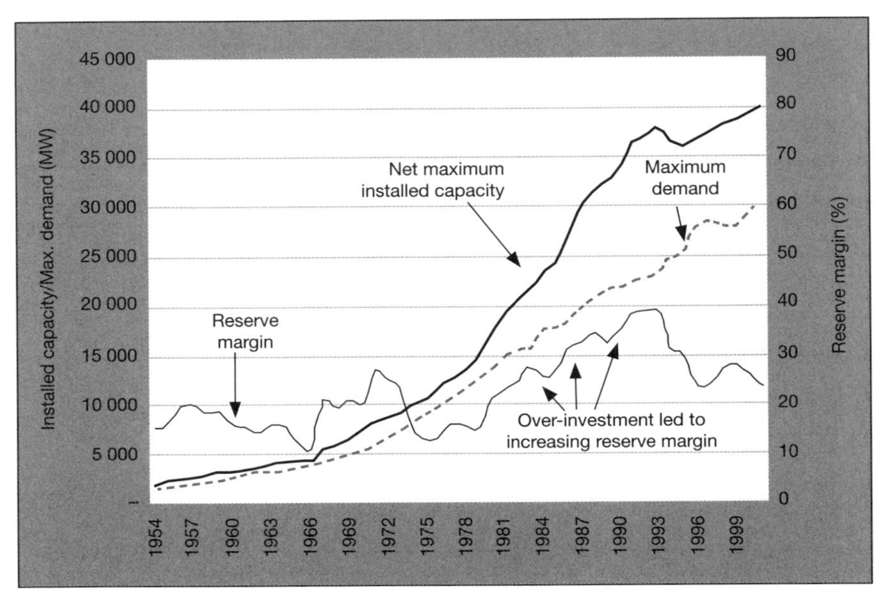

South Africa's electricity capacity and demand.

Source: Anton Eberhard, n.d.

commercial farmers during the 1980s. Since the loans were guaranteed by the state, all taxpayers, regardless of whether they benefited from the expansion of infrastructure or not, paid the bill. The World Bank's US$100 million in Eskom loans from 1951–67 and subsequent bond purchases by international banks are coming under more scrutiny as victims of apartheid seek reparations in US and European courts for the Eskom interest and profits the banks earned while black South Africans suffered.

Although industrial users do provide a small cross-subsidy to household consumers, Eskom supplies the large firms with the cheapest industrial electricity in the world. While in other countries, domestic consumers are charged twice as much as large industry, Eskom charges industry prices that are as little as one-seventh of the domestic price (Leslie 2000). As a result, the University of Cape Town's Energy for Development Research Centre (EDRC) confirms that generation of cheap electricity in South Africa still relies on the extremely wasteful burning of low-grade coal, which has a worsening impact on the environment, not only through emissions, but also in requiring vast amounts of coolant water. Indeed, Eskom is the single largest consumer of raw water in South Africa. While industry benefits from cheap electricity as a competitive advantage, the negative social and environmental effects of electricity production have never been counted into the cost.

One EDRC study concedes that South Africa:

- is 'the most vulnerable fossil fuel exporting country in the world' if the Kyoto Protocol is adopted, according to an International Energy Agency (IEA) report;
- scores extremely poorly 'on the indicators for carbon emissions per capita and energy intensity';
- has a 'heavy reliance' on energy-intensive industries;
- suffers a 'high dependence on coal for primary energy';
- offers 'low energy prices' which in part is responsible for 'poor energy efficiency of individual sectors' and
- risks developing a 'competitive disadvantage' by virtue of 'continued high energy intensity', which in the event of energy price rises 'can increase the cost of production'. (Spalding-Fecher 2000)

In short, the existing levels of environmental degradation caused by coal mining, electricity generation, a lack of access by the majority of low-income people, hydropower and nuclear energy are formidable. Not including net exports of greenhouse gas pollutants – since South Africa is the world's second-largest exporter of coal after Australia – the energy sector contributed 78 per cent to South Africa's share of global warming and more than 90 per cent of all carbon dioxide (CO_2) emissions in 1994. These ratios have probably increased since then.

By 1998, South Africa emitted 354 million metric tonnes of CO_2, equivalent to 2 291 kilograms of carbon per person (a 4 per cent increase from 1990 levels). South Africa' energy sector is amongst the worst emitters of CO_2 in the world – when corrected for both income and population size, worse than even the United States, by a factor of twenty.

South Africa took no action to reduce emissions over the period 1990–98 and indeed allowed them to increase from 2 205 to 2 291 kilograms of carbon per person (IEA 2000).

Energy sector carbon emissions, 1999.[3]

Area	Population (millions)	CO_2 per person	GDP (US$ billion)	CO_2/GDP (kg/US$ billion)	CO_2 (kg)/GDP population
South Africa	42	8.22	164	2.11	0.0501
Africa	775	1.49	569	1.28	0.0016
United States	273	20.46	8 588	0.65	0.0023
OECD	1 116	10.96	26 446	0.46	0.0004
World	5 921	3.88	32 445	0.71	0.0001

Source: IEA data, with final column calculated by Patrick Bond

Liberalisation and price sweeteners for corporations

The 1986 White Paper on Energy Policy set the framework for the marketisation of the electricity sector. It called for the 'highest measure of freedom for the operation of market forces', the involvement of the private sector, a shift to a market-oriented system with a minimum of state control and involvement, and a rational deregulation in energy pricing, marketing and production (Charles Anderson Associates 1994: 12–13). As electricity provision became increasingly politicised during the 1980s, in part because of township payment boycotts, a joint National Energy Council/Eskom workshop held in 1990 called for deregulation of the supply industry. The workshop also put forward proposals to adopt a market-oriented approach to distribution, including large, restructured distributors that would purchase power from a broker. The introduction of specific tariffs would separate generation and transmission and transmission and distribution functions (the seeds of ring-fencing). Notably, the workshop called for supply to be run on business lines (Charles Anderson Associates 1994: 15–17).

By the time of South Africa's liberation, because of heavy mining and industrial usage, per capita electricity consumption soared to a level similar to Britain, even though black South Africans were denied domestic electricity for decades. Today, most poor South Africans still rely for a large part of their lighting, cooking and heating energy needs upon paraffin (with its burn-related health risks), coal (with high levels of domestic and township-wide air pollution) and wood (with dire consequences for deforestation). Women, traditionally responsible for managing the home, are more affected by the high cost of electricity and spend greater time and energy searching for alternative energy. Ecologically sensitive energy sources, such as solar, wind and tidal, have barely begun to be explored, while the main hydropower plant that supplies South Africa from neighbouring Mozambique is based on a controversial large dam, with two others proposed for construction on the Zambezi River.

Nevertheless, Eskom claims to be one of the new South Africa's success stories, having provided electricity to more than 300 000 households each year during the 1990s. Black residents were denied Eskom's services until the early 1980s due to apartheid and the townships were, as a result, perpetually filthy because of coal and wood soot. From 1990 to the end of 2001, Eskom and the municipalities had together made nearly four million household connections, including farm workers, at a cost to Eskom of R7.72 billion (Department of Minerals and Energy 1997). The percentage of households with access to electricity infrastructure increased to 70 per cent at the end of 2000. In urban areas, the percentage of households with electricity infrastructure was 84 per cent, with rural areas lagging behind at 50 per cent (National Electricity Regulator 2001: 14).

Critics argue that regulation of Eskom and the municipal distributors has not been successful, from the standpoint of mass electricity needs (Winkler and Mavhungu 2001:

6). This is not only because of an extremely weak performance by the initial National Electricity Regulator – Xolani Mkhwanazi, who subsequently became chief operating officer for BHP Billiton Aluminium Southern Africa – but also because government policy has increasingly imposed 'cost-reflective tariffs', as a 1995 document insisted. The 1998 White Paper was an improvement on previous versions, allowing for 'moderately subsidised tariffs' for poor domestic consumers. But it too made the counterproductive argument that 'cross-subsidies should have minimal impact on the price of electricity to consumers in the productive sectors of the economy' (Department of Minerals and Energy 1998). This philosophy remained intact during Phumzile Mlambo-Ngcuka's reign as energy minister until 2005.

This raises the crucial question of the price charged to these 'productive sectors', namely a tariff regime inherited from the apartheid era that is extremely generous to minerals/metals smelters and other large electricity consumers. The man responsible for Eskom's late-apartheid pricing – Mick Davis – left the parastatal's treasury to become the London-based operating head of BHP Billiton, once former finance minister Derek Keys gave permission for Gencor to expatriate vast assets to buy the firm from Shell (after apartheid ended, Keys tellingly became chief executive of BHP Billiton).

Ten years later, the deals which gave BHP Billiton, AngloAmerican and other huge corporations the world's lowest electricity prices came under attack by Alec Erwin, minister of public enterprises. It seemed like progress finally, because the package Davis had given BHP Billiton for the Alusaf smelters at Richards Bay Hillside and Mozal in Maputo during the period of Eskom's worse overcapacity, had resulted in ridiculously cheap electricity – often below R0.06/kilowatt hour – when world aluminium prices fell. Creamer's Engineering News reported in June 2005 that, 'following the introduction of new global accounting standards, which insist on "fair value" adjustments for all so-called embedded derivatives . . . Eskom admits that the sensitivities are substantial and that the volatility it could create is cause for concern'. Erwin reportedly insisted on lower 'financial-reporting volatility' – every time the rand changes value by 10 per cent, Eskom wins or loses R2 billion – and he gave 'guidance that the utility should no longer enter into commodity-linked contracts and that management should attempt to extricate the business from the existing contracts'. Mkhwanazi replied that any change to the current contracts could be 'a bit tricky for us . . . We would adopt a pragmatic approach and, who knows, perhaps there will even be some sweeteners in it for us' (Creamer Media Engineering News 2005).

How did this new approach play out in terms of the vast subsidies promised at Coega, where Erwin as trade and industry minister from 1996–2004 had led negotiations for a new aluminium or zinc smelter? The answer was clear within two weeks, as a long-awaited US$2.5 billion (R16.3 billion) deal with Canada's Alcan came closer to completion.

According to the chief executive of the parastatal Industrial Development Corporation (IDC), Geoffrey Qhena, 'The main issue was the electricity price and that has been resolved. Alcan has put a lot of resources into this, which is why we are confident it will go ahead.' Meanwhile, however, to operate a new smelter at Coega, lubricated by at least 15 per cent IDC financing, Alcan and other large aluminium firms were in the process of shutting European plants that would produce 600 000 metric tonnes between 2006 and 2009, simply 'in search of cheaper power', according to industry analysts.

If it is eventually built (a matter less likely in 2008, thanks to national electricity shortfalls and Alec Erwin's departure from cabinet), a Coega plant would generate an estimated 660 000 tonnes of CO_2 a year. For the purpose of complying with Kyoto Protocol obligations, Europe will be able to show reductions in CO_2 associated with the vast energy intake needed – representing one-third of a typical smelter's production costs – while South Africa's CO_2 will increase proportionally. Indeed, as a result of the sweeteners offered to Alcan, Eskom will run out of its excess electricity capacity more rapidly, resulting in raised prices to poor people, more coal generation and a more rapid turn to objectionable power sources, such as nuclear reactors and the two proposed Zambezi River megadams (Bailey 2005).[4]

According to Richard Fuggle, a professor at the University of Cape Town:

> It is rather pathetic that our current environment minister Marthinus van Schalkwyk has expounded the virtues of South Africa's 13 small projects to garner carbon credits under the Kyoto Protocol's Clean Development Mechanism, but has not expressed dismay at Eskom selling 1360 megawatts a year of coal-derived electricity to a foreign aluminium company. We already have one of the world's highest rates of carbon emissions per dollar of GDP. Adding the carbon that will be emitted to supply power to this single factory will make us number one on this dubious league table. (2006)

Price hikes and disconnections for the poor

The contrast with the government's treatment of low-income people is stark. While Eskom was offering billions of rand's worth of 'sweeteners' to the aluminium industry, the Department of Provincial and Local Government's 'Municipal Infrastructure Investment Framework' supported only the installation of 5–8 amp connections for households with less than R800 per month income, which does not offer enough power to turn on a hotplate or a single-element heater (in turn, without a higher amperage, the health and environmental benefits that would flow from clean electricity instead go up in smoke). The 1995 energy policy also argued that 'fuelwood is likely to remain the primary source of energy in the rural areas'. Eskom did not even envisage electrifying the nation's far-flung schools, because 'it is not clear that having electricity in all schools is a first priority' (Department of Minerals and Energy 1995).

Moreover, Eskom economists had badly miscalculated rural affordability during the late 1990s, so revenues were far lower than were considered financially sustainable. Because of high prices, consumption of even those households with five years of access was less than 10 kilowatt hours per month, resulting in enormous losses for Eskom. Paying as much as R0.40 per hour (compared to a corporate average of R0.06 and bigger discounts for Alusaf), rural women used up their prepaid meter cards within a week and can't afford to buy another until the next pension payout. This was the main reason demand levels are so low that Eskom's rate of new rural electrification connections ground to a standstill.[5]

The state's electricity subsidy was insufficient to make up the difference, even when the African National Congress (ANC) government introduced its free basic services policy in mid-2001. Eskom refused to participate for several years, waited until a new national subsidy grant became available and still today has not fully rolled out the promised 50 kilowatt hours per household per month lifeline supply. In 2008, Eskom told low-income households to revert to paraffin stoves.

Politicians and municipal managers defend the system, notwithstanding these many problems. The leading official of eThekwini (Durban), Mike Sutcliffe, justifies the inadequate 50 kilowatt hours/household/month allocation:

> The amount of 50 kWh was developed at national level in consultation with Eskom where 56 per cent of their residential customer base currently uses less than 50 kWh a month and this includes many customers in colder climates than Durban. The average consumption of all our prepayment customers (160 000) is 150 kWh a month and not all of them are indigent.
>
> South Africa does not have sufficient experience in the provision of free energy services to conclude whether 50 kWh a month is adequate or not. The amount of 50kWh would appear to be a reasonable level to start with on a nationwide basis using the self targeted approach. If the self targeting works and the country can afford to increase the free service it could be reviewed in the future. There are more than 7 million electrified households in South Africa and for every 1 million indigent households receiving 50 kWh free the loss in revenue is R17,5 million a month.
>
> The proposal of a flat rate has proven to result in considerable wasted energy as users are unaware of their usage and consume far more than that

> which could be purchased for R50. Even if a current limit of only 10A is imposed these flat rate users could consume well over 1 000 kWh a month. South Africa can ill afford to waste energy, the generation of which not only depletes our fossil fuel reserves but has a considerable impact on water resources used in the generation process and air pollution as 80 per cent of South Africa's generation is from coal. (Sutcliffe 2003)

Whether or not Sutcliffe has resorted to data torture to get the statistics to confess, it is not at all unusual for wealthy South Africans – perhaps suffering from a culture of privilege – to advocate that poor people should consume less electricity or water because they 'waste' these state services (it may be irrelevant, but Sutcliffe earns a far greater income than the South African president). Uniquely, though, Sutcliffe here also implies the poor are responsible for depleting the vast South African coal reserves, even though household electricity consumption by low-income families in South Africa is still less than 5 per cent of the national total.

Misleading or wildly inaccurate information from state officials – relating to, for example, AIDS, arms deals, crime, adult education and municipal services – is an epidemic in South Africa, a country also overpopulated by gullible journalists. Witness a South African Press Association (SAPA) report, reprinted in the *Mail & Guardian* in March 2005 about a services survey by Statistics South Africa (the government's official statistical service): 'The best-performing municipalities on average were in the Free State, where 91.5 per cent of households had free water and 90.3 per cent had free electricity'. The explosive municipal riots in the Free State must indeed have been a right-wing plot, as alleged by some in the ANC, since denial of services was obviously not a factor. Conveniently, it would apparently be impossible to verify these amazing claims, because 'Stats SA said although it is able to release provincial data, it cannot in terms of the Statistics Act release unit information – that of individual municipalities in this case – without their express permission'. According to Statistics South Africa head Paddy Lehohla: 'Municipalities do need to be protected by the Act because they may want to apply to certain organisations for grants, and poor performance figures could harm them, or there may arise situations where they face punitive measures from the ruling party in their areas' (Mahlangu 2005).

For very different reasons, some in national government periodically concede that low-income South Africans do not, in fact, receive sufficient free electricity. In November 2004, prior to taking over as deputy-president from Jacob Zuma, energy minister Pumzile Mlambo-Ngcuka alleged, according to the South African Broadcasting Corporation (SABC), that 'municipalities are botching up government's free basic electricity initiative to the poor'. The SABC report continued: 'However, there is another bureaucratic dimension to the problem. Eskom, a state-owned enterprise, is struggling to recoup its money from the Treasury for the free electricity it provides and Mlambo-Ngcuka says even when Eskom does get the money from them, it is always insufficient.' Indeed, the Treasury's 2004 grant of just R200 million to cover free basic electrification subsidisation is grossly inadequate. But Mlambo-Ngcuka's own ministry was mainly to blame. Her staff had obviously overruled the 2000 ANC election promise of free basic services through a rising block tariff, for they apparently remained committed, instead, to 'cost-reflective' pricing of electricity (not counting the sweetener deals with the aluminium industry).[6]

In a similar vein, when the World Bank came under pressure in 2004 for its sweet financing of extractive industries, Mlambo-Ngcuka again revealed her loyalties, making it clear to senior Bank staff in February 2004 that they should oppose 'green lobbyists', as reported by the UN news agency Integrated Regional Information Networks (IRIN). Instead of the Extractive Industries Review (EIR) provisions for a phase-out of Bank fossil-fuel investments, Mlambo-Ngcuka promoted the African Mining Partnership within the neoliberal New Partnership for Africa's Development (NEPAD). According to her spokesperson, 'We are already implementing sustainable development programmes'.[7]

The energy system Mlambo-Ngcuka and her replacement Buyelwa Sonjica oversaw was anything but sustainable for its many victims. By pricing electricity out of reach of the poor, the state officials, economists and consultants who design tariffs together refuse to recognise multiplier effects that would benefit broader society, were people granted a sufficient free lifeline electricity supply. One indication of the health implications of electricity supply disconnections that resulted from overpriced power was the recent upsurge in tuberculosis rates. Even in communities with electricity, the cost of electricity for cooking is so high that, for example, only a small proportion of Sowetans with access to electricity use it, favouring cheaper fuels.[8] The gender and environmental implications are obvious.

The result of unaffordable electricity and inadequate state subsidies was an epidemic of disconnections. Electricity cut-offs were widespread by 2001. At that point, the Department of Provincial and Local Government's project viability reports and Eskom press statements together indicated an electricity disconnection rate of around 120 000 households per month. These are likely to be higher, since not all municipalities responded to the Department's survey and the Eskom statements focus on Soweto, where resistance was toughest. But even using this base and making a conservative estimate of six people affected by every disconnection (since connections are made to households that often have tenants and backyard dwellings) upwards of 720 000 people a month were being disconnected from their access to electricity due to non-payment, meaning that there were several times as many households losing access to electricity every month, as were gaining access. A survey of Soweto residents found that 61 per cent of households had experienced electricity disconnections, of whom 45 per cent had been cut off for more than one month. A random, stratified national survey conducted by the Municipal Services Project and Human Sciences Research Council (HSRC) found ten million people across South Africa suffered electricity cuts (McDonald 2002).[9]

Even higher numbers could be derived using municipal disconnection statistics available through project viability, a national accounting of municipal finances whose last data set was analysed by the Department of Provincial and Local Government in December 2001. After that date, the embarrassing statistics have not been publicly available, in spite of numerous requests by Centre for Civil Society (CCS) students. The latest report showed that 174 municipalities out of a total of 284 implemented credit control procedures that included service disconnections. During the last quarter of 2001, those 174 municipalities disconnected electricity to 296 325 households due to non-payment. Of those, 152 291 households were able to pay a sufficient amount to assure reconnection during the quarter, leaving 144 034 families – 4.3 per cent of the total population connected – without electricity at Christmas in 2001. If, very conservatively, half a million people were adversely affected

Latest available project viability statistics (October to December 2001).

The number of councils that	
have formally approved credit control and procedures	174
do credit checks before opening an account	34
read the meters for water	
– monthly	185
– every 2–3 months	2
– irregularly	2
– never	93
read the meters for electricity	
– monthly	154
– every 2–3 months	2
– irregularly	2
– never	93
render consumer accounts to all areas in the jurisdiction	159
The number of	
electricity disconnections done over the past 3 months	296 325
electricity reconnections done over the past 3 months	152 291
water disconnections done over the past 3 months	133 456
water reconnections done over the past 3 months	50 703
summons issued over past 3 months	58 498
The number of households	
receiving water	4 084 009
receiving electricity	3 366 226

Source: Department of Provincial and Local Government 2002: 30–31

during this quarter – a time when December bonuses should have permitted bill arrears payments –then, multiplying by four quarters, roughly two million people would, cumulatively, have had their power disconnected for substantial periods (on average 45 days) throughout 2001.

Moreover, since Eskom supplies more than half the low-income township population directly, and since self-disconnecting prepaid metered accounts are not included in these statistics, the number of people who lost power would logically be far higher. Hence the electricity attrition rate – i.e., the percentage of those who were once supplied with electricity, but who could not afford the high prices and lost access due to disconnections – must be, using these indicative statistics, scandalously high for South Africa as a whole. Indeed, the ongoing lack of electricity supply to low-income people is invariably blamed, in part, for the upsurge in municipal protests since the early 2000s.

Rising electricity prices across South African townships already had a negative impact during the late 1990s, evident in the declining use of electricity, despite an increase in the number of connections. According to Statistics South Africa, households using electricity for lighting increased from 63.5 per cent in 1995 to 69.8 per cent in 1999. However, households using electricity for cooking declined from 55.4 per cent to 53.0 per cent from 1995 to 1999 and households using electricity for heating dropped from 53.8 per cent in 1995 to just 48.0 per cent in 1999.

Although comparable data are not available for the subsequent five years, in 2001 Statistics South Africa conceded a significant link between decreasing usage and the increasing price of electricity and there is no reason to believe that this trend was subsequently reversed (Statistics South Africa 2001: 78–90). The implications for women and children are most adverse, given the inhalation of particulates that they suffer during cooking and heating with coal, wood or paraffin.

The implications for social unrest cannot yet be quantified, although the South African Police Service has recorded the number of Gatherings Act protests at nearly 9 000/year over the three-year period 2004–07.

The renewable energy funding drought

In contrast to the vast amounts of energy generated through dirty coal-fired methods, South Africa's renewable sources with enormous potential include solar and wind, but these are surprisingly underdeveloped, as Muna Lakhani and Vanessa Black note in Chapter 2. Capital costs are expensive, as are repairs.

Resource allocation by the South African government remains skewed away from renewable energy, towards nuclear research and development. In 1995–96, energy spending through the Department of Minerals and Energy was R515 million, of which R489 million went to the Atomic Energy Corporation (AEC), mainly for debt servicing, even though the AEC produced no new electricity since nuclear power generation had been purchased by Eskom. In addition, that year, the Central Energy Fund (CEF) wrote down loans to Soekor by more than R110 million and included additional provisions for non-payment of loans to state companies by R7.3 billion. Another R1.5 billion was spent on subsidising synthetic fuels. Eskom's capital investments that year amounted to R5.4 billion and there were many other unaccounted investments in energy, through local electricity distributors, transport or pipeline companies, state oil companies, Eskom and National Research Foundation research and development in energy and upgrading of port infrastructure for coal handling. The problem of resource allocation appears to be getting worse. Expenditure on renewable energy was less than 0.5 per cent of the Department of Minerals and Energy's budget in 2002–03.

Ironically, this was the moment that the South African government released its 'White Paper on Renewable Energy', which claims that electricity generation from renewables will reach 4 per cent by 2013. As Graham Erion shows elsewhere in this volume, however, the statistic is misleading: 'For starters, the 4 per cent target is *cumulative*, meaning that it will be satisfied if the annual percentage of electricity coming from renewables every year adds up to 4 per cent by 2013. Therefore if new renewable capacity goes online next year totally just 0.5 per cent of the market and no other new supply goes online, this target will be satisfied.'

The nuclear funding flood

Meanwhile, the Department of Minerals and Energy continued to fund the Nuclear Energy Corporation of South Africa (NECSA; successor to the AEC), in 2001, to the tune of R135 million plus strategic loans of R266 million (Department of Finance 2001: 706–09). Opposition to nuclear energy on grounds of safety and long-term waste storage has come from various sections of civil society, notably most of the environmental movement and the trade unions. By the end of 2001, the Congress of South African Trade Unions (COSATU) and four dozen other civil society organisations and networks were joined by another 23 regional and international organisations in opposition to a nuclear development path in South Africa (Earthlife Africa 2002; see also 2001a, 2001b).

By November 2004, Earthlife Africa had won a court battle against Chippy Olver, former director-general of the Department of Environmental Affairs and Tourism, for his failure to take into account their views during nuclear energy environmental impact hearings. In January 2005, Olver was forced to turn over files he had refused to give Earthlife regarding the nuclear programme's safety, complaining of 'a seemingly endless round of consultations and judicial reviews' (I-Net Bridge 2005). A related controversy emerged in October 2005, when the president of the World Conservation Union (IUCN), former environment minister Vali Moosa (who oversaw Olver's decisions), had to defend his mid-2005 acceptance of the chairmanship of Eskom to IUCN board members in Geneva, who were aghast at the corporation's environmental record.

From an economic point of view, the cost of production of the preferred nuclear option – the pebble bed modular reactor (PBMR), which is 50 per cent owned by Eskom – became unviable during the early 2000s, given currency fluctuations and severe problems experienced by Eskom's partners in Britain and the United States. On a simple (non-environmental) financial basis, electricity generated from nuclear power in other countries costs up to 25 per cent more than conventional fuels (Earthlife Africa 2001c). The PBMR technology had been rejected by German firms who sold it to Eskom, yet is presently being marketed as home-grown South African knowledge because of an as yet unproven alteration to the design.

In spite of the vast waste of resources, the nuclear programme has been expanded during the post-apartheid era, as pointed out in the next chapter. Against all evidence to the contrary, such as the departure of US investor Exelon, public enterprises minister Alec Erwin claimed to parliament in October 2004: 'There are constant requests for information from different governments, utilities and research institutions on the PBMR technology'. Asked about the costs to taxpayers, Erwin replied in manner that has become familiar: 'Given that there are other shareholders involved, and the project is in a fund-raising exercise, this information is confidential and cannot be divulged' (SAPA 2004b). The fundraising failure became obvious a few months later, when Trevor Manuel authorised dropping another R500 million from the fiscus into the PBMR sinkhole. According to an e-mail from Earthlife campaigner, Sibusiso Mimi on 23 February 2005:

> The project is moving backwards. The projection by Phumzile Mlambo-Ngcuka in her budget speech, saying nuclear energy is inevitable for South Africa, and that by 2010 the PBMR will be economically viable, is a lie. Eskom has just announced another verdict: that the PBMR will be economically viable by 2013. In essence, the project has moved three more years backward in few months after the budget speech despite a generous R500 million, which really means that South Africa is being used as a testing ground for this white elephant alienated by the global investing community, while its proponents are praising it like some kind of a god.

Earthlife's protest was joined by the South African Council of Churches, South African Non-Governmental Coalition and COSATU:

> Government intends allocating R500-million to the PBMR. At the same time, government has allocated slightly more than a billion rand in the 2004/05 financial year for the national electrification programme. The spending on the PBMR is almost half of the projected spending to achieve universal access. The project involves high risks and unpredictably high costs with the prospect of limited returns. (I-Net Bridge 2005)

The most recent critique of PBMR, from Greenwich University researchers, was covered by the *Cape Times* in August 2005:

> South Africa will have to spend a massive R25-billion on the proposed pebble bed nuclear power project before it will be economically viable. This has emerged from an international report on the economic impact of the proposed pebble bed modular reactor which says that if the project goes ahead South African consumers could end up paying for 'a series of expensive white elephants.'
>
> The cost of a PBMR demonstration plant to be built at Koeberg has risen from R2-billion in 1999 to R14-billion today. This excludes the decommissioning costs, which would be at least another R5-billion. The economic forecasts by PBMR are 'implausibly optimistic.'
>
> The economic report was written by Steve Thomas, of the Public Service International Research Unit at the University of Greenwich and commissioned by the Legal Resources Centre. It is to form part of a submission by Earthlife Africa to the department of environment affairs.
>
> The department was ordered by the Cape High Court six months ago to reopen the environmental impact process for the pebble bed, but has not yet done so. The National Environmental Management Act requires that the state ensure development is economically sustainable. Thomas writes that South Africa plans to build several of the nukes for export but, after years of negotiations, has no overseas orders.
>
> The developer, PBMR, is pressuring Eskom to commit, unconditionally, to buying 24 of the units at a cost of R25-billion. This would allow 'economies of scale' to kick in and only then could the company produce a commercially competitive product.
>
> Thomas says the PBMR's huge escalating costs and the long time delays show that the developers have failed to understand the nature or scale of their

task. Their poor track record gives little confidence that they would be able to control costs and time schedules in the next, more expensive, phase.

The pebble bed's economic forecasts by the PBMR company have not been updated since 1998 and are 'implausibly optimistic.' Thomas points out that, as the demonstration plant itself would only incur costs, not create profits, building it would make sense only if there were a high probability of a 'stream of orders' from overseas.

Beijing has made no commitment to buy PBMRs. The company had been 'very vague' about its target markets. Its analysis of the world nuclear market was simplistic and its assumptions about who would buy the exported PBMRs had no basis. There was 'nothing remotely close to a firm order' from overseas for a pebble bed nuke reactor. The main expected export market was China but, despite several years of discussions, Beijing had made no commitment.

South Africa has not been able to find another international partner for the nuke project since the US company, Exelon, pulled out in 2002. John Rowe, chief executive officer of Exelon, said the reason for the withdrawal was that 'the project was three years behind schedule and was too speculative.'

The French nuclear company Areva has also indicated it is not prepared to fund the demo plant. Britain's BNFL, the only foreign partner, is in financial difficulties.

Thomas says the PBMR project has always been high-risk and the risks were likely to fall squarely on the shoulders of the South African public. As South Africans would have to be the major underwriters for the pebble bed project, it was 'reprehensible' that most of the economic information needed to evaluate it had been withheld from the public. 'It is particularly regrettable that a report by an international panel of experts, commissioned by the department of minerals and energy to review the overall project, has not been made public', Thomas wrote.

Thomas, a member of the panel, said the panel had been 'required to promise not to disclose any information' about the report. The Legal Resources Centre has tried, under the Access to Information Act, to get the department of minerals and energy to release the report, but it has refused to do so.

Peter Bradford, former commissioner of the US Nuclear Regulatory Commission, peer-reviewed Thomas's report this month and his only criticism was that Thomas had been 'conservative' in his concerns about the pebble bed. Bradford said Thomas had not considered the negative impact on the

> South African economy that would flow from electricity bill increases or tax increases to fund the pebble bed project. He also had not considered that the Chinese pebble bed design or the Areva prismatic nuclear design were likely to be effective competitors for whatever market developed for the pebble beds. (Gosling 2005)

Clearly the multiple environmental, social and economic dangers posed by the government's nuclear fetish are substantial. Sensitivities at the highest levels of government are one indication that Earthlife is on the right track. After the organisation – of which I confess to be an ordinary Durban branch member – revealed high levels of radioactivity near the Pelindaba plant, politicians went ballistic. Energy minister (later deputy-president) Mlambo-Ngcuka warned: 'We are considering strengthening the law so that if people make such allegations there is a sanction'. President Mbeki, who was that weekend awarded the ludicrous UN 'Champion of the Earth' award, accused Earthlife of making 'reckless statements', which were 'in [his] view, totally impermissible . . . We cannot go on scaring people about something that does not exist . . . These statements have been made by an NGO [non-governmental organisation] in order to promote its own interests, which is regrettable.' (SAPA 2005)

However, Earthlife's agenda was a bit broader than that, as two journalists learned to their surprise:

> When the *Pretoria News* visited the site yesterday radioactive warning signs were, at first, nowhere to be seen. A chicken-wire fence had been erected around the site. Less than 30 minutes after arriving there, Nuclear Energy Corporation of South Africa (NECSA) officials 'escorted' reporters off the site before erecting 'private property' signs as well as signs warning of radioactivity.
>
> Officials from the National Nuclear Regulator (NNR) and NECSA also spent the day conducting radiation level readings. NECSA spokesman Nomsa Sithole said the signs 'are part of the organisation's security measures and are used to warn people to keep off the land.'
>
> 'I categorically deny that the site is a nuclear waste dump. All our waste is dumped within the nuclear facility itself', she said.
>
> Sithole said the site, a former calibration facility established in 1979, was used to calibrate the instruments used by Pelindaba staff. 'While I admit that the fence around the area is not up to scratch, there is no need for fear of radiation leaking from the site', she said.
>
> Sithole said the radiation warning signs had been posted to warn people about enhanced levels of 'naturally-occurring' radioactive materials mixed

> into the concrete calibration pads. She could not say why they were erected only yesterday.
>
> NNR communication manager Phil Nkhwashu confirmed they were investigating the site, but declined to comment further.
>
> Dr Stefan Cramer, a geologist who conducted tests at the site on Saturday on behalf of Earthlife, said there had been a grave lack of security and an oversight by NECSA concerning the nuclear facility. He said he had not seen such high levels of radiation in such an open area before. He claimed that the radiation in the immediate vicinity of the site was 200 times higher than natural radiation . . .
>
> Government spokesman Joel Netshitenzhe said that given the undue panic generated by the scare, South Africans, including the media, needed to be cautious when handling information from organisations 'with their own narrow agendas'. (Hosken and Adams 2005)

Earthlife Johannesburg's leading anti-nuclear campaigner Mashile Phalane explained that 'agenda': 'We want government to regulate the industry properly and punish anyone who transgresses the law' (*Business Day*, 29 April 2005). Fortunately, Earthlife will continue raising concerns about nuclear safety and other 'narrow agendas', while battling Mbeki, Mlambo-Ngcuka, Erwin, Netshitenzhe and their successors. In addition, thankfully, the Kyoto Protocol still prohibits the use of nuclear energy as justification for reducing greenhouse gases and rejected the nuclear option within the Clean Development Mechanism (CDM), but this stance is under attack from the United States and other pro-nuclear governments (Earthlife Africa 2001c: 4). It remains to be seen whether the IUCN (under Vali Moosa's presidency) advances Eskom's pro-nuclear agenda in the carbon markets, where Moosa's interests are potentially lucrative in the event the PBMR is ever built and authorised as a CDM.

The extent of nuclear ineptitude within the state was unveiled in 2006, in semantic squabbling in the Western Cape, in the wake of yet another electricity outage. On the day before a crucial municipal election in which the ruling party lost control of Cape Town, Erwin blamed a Koeberg nuclear power station shutdown on a loose bolt in a turbine and two days later incorrectly claimed, 'I did not use the term sabotage' (Pressly 2006).[10] Quipped Western Cape regional secretary of COSATU, Tony Ehrenreich: 'Regarding the power cuts, we should perhaps blame a few neoliberal nuts rather than a bolt that fell into a Koeberg generator' (Bell 2006).

When a report on the incident emerged from the National Energy Regulator of South Africa (NERSA) in August 2006, its contents were chilling enough to turn the late *Mail & Guardian* satirist Robert Kirby serious:

> . . . Eskom's lamentable security, inadequate protection systems, slack maintenance, breaches of licences, negligence and generally slovenly management at Koeberg – all of which resulted in a series of power blackouts in the Western Cape earlier this year. Never mind the blackouts, what about the now conspicuous danger of another Chernobyl disaster just up the road from Cape Town? Koeberg is of the same vintage as Chernobyl and the latter went out of control because of exactly what NERSA recently diagnosed at Koeberg: lamentable security, inadequate protection systems, et cetera. You name them, Chernobyl had them.
>
> Last week, Erwin proclaimed to Parliament on this appalling state of affairs. The failures at Koeberg were 'being dealt with through internal disciplinary procedures', he said in a prepared statement. In other words, what should be examined and analysed under the floodlights of public scrutiny, is now being held in secret.
>
> Getting curiouser and curiouser, Erwin once again denied that he ever used the word 'sabotage' when discussing the bolt in the generator. This time he abandoned his risible 'human instrumentality' and fell back instead on low specific-gravity bilge. 'It was also why I did not use the word sabotage, as we had to verify the existence or otherwise of a group before any such word was appropriate. The non-existence of such group has now been conclusively established . . .'
>
> In a prepared statement, the composition of which has taken only five months to hatch, further clarification came from Erwin himself. It included an even more woeful alibi: 'Of as much interest has been whether I said this was an act of sabotage. I did not say this and all attempts I made to our erudite media to say what I did say merely got me into deeper linguistic difficulties.'
>
> Judging only by the intellectual tensile strength of his logic, I would say that papers like *The People* or *The Sun* probably register as erudite on the Erwin scale. Anyway, in getting himself into linguistic difficulties, Alec Erwin has never needed much help from the media.
>
> Erwin denied to Parliament that he had used the word 'sabotage.' I've suddenly realised why they are letting him get away with it. Coming from Alec, the denial was just a token white lie. (Kirby 2006)

A final question came from Earthlife Africa, Cape Town, the group most responsible for maintaining public pressure: 'How many mistakes before adequate responsive action is taken and Koeberg's license is suspended?'

> Eskom has failed to respond to warnings and has failed to test vital equipment for periods of up to ten years. That Eskom should seek to conceal these problems is unacceptable and raises concerns about the parastatal's ability to remain transparent and accountable. Such institutions must ensure compliance with the law, among which is the constitutional right to a healthy, safe and clean environment as well as the information that can help us to achieve a healthy, safe and clean environment. Safety at a nuclear power station is of the highest concern. The potential for accidents at nuclear power stations and the consequences of such accidents have been well illustrated by historic events at Chernobyl and in the USA and Japan. These accidents have resulted in numerous deaths and environmental destruction as well as a legacy of radioactivity that is responsible for illnesses and deaths until today. It is shocking that this information is coming to light only eight months subsequent to the onset of problems at Koeberg. It appears further as if these failures have yet to be addressed. Even more alarming is that it has taken NERSA and not the National Nuclear Regulator to make these discoveries. (Aberman 2006)

The hydro option

There are also enormous environmental and social problems associated with hydroelectricity across southern Africa, not least of which are global warming gases released in tropical dams due to vegetation decay.[11] Favouring hydropower and the privatisation of Africa's existing energy agencies, Eskom had ventured into the following countries by 2000: Angola, Botswana, Cameroon, the Democratic Republic of Congo (DRC), Ghana, Mali, Mozambique, Swaziland, Tanzania and Zambia. In the early 2000s, major Eskom Enterprise projects included:

- a R100-million agreement to supply water and electricity in Gambia;
- a fifteen-year operation and maintenance contract for the new Manantali hydro-station in Mali and its associated high voltage transmission system;
- the formation of a consortium with the French firms EDF and Saur International to bid for 51 per cent of Cameroon's Sonel;
- an alliance agreement with the Libyan power utility, Gecol;

- an agreement with Nigeria's National Electricity Power Authority covering generation and operations, electro-mechanical repairs, transmission, and rehabilitate, operate and transfer (ROT) schemes;
- consulting and management contracts in Malawi; and even
- a bid for power stations operated by the Zimbabwean Electricity Supply Authority as repayment for outstanding debt owed to Eskom. (Greenberg 2002)[12]

There have been many other feelers in Africa recently, including major contracts in Nigeria and at Uganda's extremely controversial Bujagali Dam. As John Daniel and Jessica Lutchman of the HSRC explain:

> Hydro-electric power is regarded as a more viable option for South Africa at present and it is in this context that South Africa's developing trade and other ties to Africa loom large. Mozambique possesses substantial hydro-electric capacity (sourced from Cahora Bassa), some of which it sells to South Africa. South Africa's largest initiative is the Grand Inga project in the DRC. Grand Inga is expected to generate 40 000 Megawatts of electricity, sufficient to meet the needs of the entire continent as well as generate revenue for its members by exporting its surplus power to Europe. Grand Inga is the vital element in South Africa's long-term objective of ensuring its self-sufficiency in electricity. It is little wonder then that the South African government has committed so much in the way of time and effort, as well as military peacekeepers, to the task of bringing political stability to the DRC. (Daniel and Lutchman 2005)

The danger of this sort of hydropower hype is obvious, however, and was recognised in 1998–2001 World Commission on Dams studies of large energy and irrigation facilities associated with megadams, which invariably failed to meet economic expectations. As International Rivers Network campaigner Terri Hathaway put it in a useful corrective, reliance upon Inga may not be advisable given

> Africa's vulnerability to climate change and political instability. Climate change will bring risks to hydro-dependent economies through increases in the severity and frequency of both droughts and floods. Worsening droughts will reduce hydropower production, while increased floods threaten dam safety and may also increase sedimentation (thus shortening the useful life of dams). Climate change will add to existing environmental stresses on riverine eco-

> systems and watersheds. Economic feasibility, environmental impact studies and engineering plans for Inga should take into account the hydrological uncertainties of a warming world.
>
> Political instability is a very real concern across the region where the transmission grid would be built. The ongoing violence in DRC was recently rated the world's most forgotten crisis by Reuters. Over three million people have died since 1998 as a result of the civil war and ongoing strife in DRC. The Inga mega-project would centralise much of Africa's electricity source and require a grid of transmission lines through many of Africa's most politically unstable regions. Dams, power plants, and transmission lines are often made targets in political conflicts. The dependence of more countries' economies on Inga would increase its attractiveness as a target for sabotage by rebel groups. Less than 10 years ago (in 1998), rebels seized Inga II and cut its power to Kinshasa, the capital of the DRC. (Hathaway 2005)

Economic and political corruption

Two other contextual factors in South African energy must also be addressed so as to provide sufficient background on why the climate strategy is doomed: first, the disastrous impact of the Washington Consensus on economics and second, crony capitalism.

As noted in the Introduction, there is unending self-congratulation about the few economic variables that have turned positive in recent years, in the wake of South Africa's longest ever depression, from 1989–93. Specifically, as finance minister Trevor Manuel bragged in 2006, there is an oft-repeated claim that the ANC has created 'a level of macroeconomic stability not seen in the economy in decades'. Several points should be raised in rebuttal:

- The value of the rand crashed by more than a quarter in 1996, 1998, 2001 and 2006, the worst record of any major currency, which in turn reflects how vulnerable South Africa has become to whimsical international financial markets, thanks to steady exchange control liberalisation that started in 1995.
- South Africa has witnessed GDP growth since 1999, but this does not take into account the depletion of non-renewable resources; if this factor plus pollution were considered, South Africa would have a net *negative* per person rate of national wealth accumulation, even according to the World Bank (as we see in Chapter 8).
- South Africa's economy has become much more oriented to profit-taking from financial markets than production of real products, in part because of 'sado-monetarism'. From March 1995 (when the finrand exchange control was relaxed, allowing capital to flood

out of South Africa), the after-inflation interest rate rose to a record high for a decade's experience in South African economic history, often reaching double digits. After a recent 3.5 per cent spike, consumer and housing credit markets are badly strained by serious arrears and defaults.

- The two most successful major sectors from 1994–2004 were communications (12.2 per cent growth per year) and finance (7.6 per cent) while labour-intensive sectors such as textiles, footwear and gold mining shrunk by 1–5 per cent per year, and overall, manufacturing as a percentage of GDP also declined.
- Government admits that overall employment growth was −0.2 per cent per year from 1994–2004, but −0.2 per cent is a vast underestimate of the problem, given that the definition of employment includes such work as 'begging' and 'hunting wild animals for food' and 'growing own food', according to Statistics South Africa.
- The problem of excessive capital intensity in production – too many machines per worker – will probably get worse. South Africa's IDC (a state agency) forecasts that the sector with the most investment in the period 2006–10 will be iron and steel, with a massive 24 per cent rise in fixed investment per year. But this sector is also one of the leading sectors for job-shedding, with employment expected to fall 1.3 per cent per year, in spite of – or indeed because of – all the new investment.
- Overall, the problem of 'capital strike' – the big business failure to invest – continues, as gross fixed capital formation hovered between 15–17 per cent from 1994–2004, hardly enough to cover wear and tear on equipment, according to the South African Reserve Bank.
- Where did businesses invest if not in South Africa? Dating from the time of political and economic liberalisation, most of the largest Johannesburg Stock Exchange firms – AngloAmerican, De Beers, Old Mutual, SA Breweries, Liberty Life, Gencor (now the core of BHP Billiton), Didata (Dimension Data), Mondi and others – shifted their funding flows and even their primary share listings to overseas stock markets.
- The outflow of profits and dividends due these firms is one of two crucial reasons South Africa's 'current account deficit' has soared to among the highest in the world – at 9 per cent of GDP in mid-2008 – and is hence a major danger in the event of currency instability, as was Thailand's (around 5 per cent) in mid-1997.
- The other cause of the current account deficit is the negative trade balance. Blame this on a vast inflow of imports after trade liberalisation, which export growth cannot keep up with.
- Another reason for capital strike is South Africa's sustained overproduction problem in existing (highly monopolised) industry. Manufacturing capacity utilisation fell

substantially from the mid-80 per cent range during the 1970s, to the high 70 per cent range during the early 2000s, according to the South African Reserve Bank.

- So where did corporate profits go, if not into over-accumulated plant, equipment and factories? The answer is obvious: speculative real estate and the Johannesburg Stock Exchange. There was a 50 per cent increase in share prices during the first half of the 2000s. The property boom began in 1999 and by 2004 house prices had risen by 300 per cent, in comparison with just 60 per cent in the US market prior to the burst bubble, according to the IMF.
- In sum, is this 'macroeconomic stability'? Or instead, a parasitical, slow-growth, high-poverty, unemployment-ridden, more unequal, capital-flight-prone, volatile, vulnerable, elite-oriented economic machine ploughing over poor people, whose main gains are in temporarily restored profitability for big capital and conspicuous consumption bingeing for a credit-drowned petit bourgeoisie?

If the macroeconomic fundamentals are shaky, what about the political context? At the end of 2007, well-known journalist Sam Sole summed up matters:

> What a country. Both our president-in-waiting and our police chief separately face the prospect of corruption and racketeering charges; our previous national director of public prosecutions was accused of once being an apartheid-era spy and all but hounded out of office for pursuing the first investigation; our current national director was suspended by the president for pursuing the second; our intelligence service boss is put on trial for playing politics and accuses his political masters of doing the same; our dominant party raises R907-million in funding over five years, but won't say where from and will not disclose the details of its R1,75-billion worth of investments. South Africa's recent history is hardly comprehensible without deconstructing the impact of covert networks of influence that find expression in the overlap of state intelligence, political funding and organised crime. (Sole 2007)

Conclusion

In sum, several important factors converge when we consider the nature of South African energy:

- South Africa, already among the most unequal countries in the world in 1994, became more unequal during the late 1990s, as a million jobs were lost due largely to the stagnant economy, the flood of imports and capital or energy-intensive investment – and these trends had enormously negative implications for the ability of low-income citizens to afford electricity;

- billions of rand in state subsidies are spent on capital-intensive energy-related investments such as new smelters, where profit and dividend outflows continue to adversely affect the currency;
- the price of electricity charged to mining and smelter operations is the lowest in the world;
- a pittance is being spent on renewable energy research and development, especially compared to a dubious nuclear programme;
- the dangers of nuclear energy are now widely understood, in the wake of damaging reports on the Koeberg power plant showing systemic maintenance problems that should result in the plant's decommissioning;
- greenhouse gas emissions per person, corrected for income, are among the most damaging in the world and have grown worse since liberation;
- electricity coverage is uneven and notwithstanding a significant expansion of coverage, millions of people have had their electricity supplies cut as the state provider moves towards commercialisation and privatisation;
- the possibilities of improving gender equity through access to free lifeline electricity are vast but unrealised;
- for people suffering from the recent upsurge in tuberculosis and indeed for 6.4 million HIV-positive South Africans, the public and personal health benefits of replacing coal, wood or paraffin with electricity are also vast and also unrealised; and
- there are other important environmental, segregation-related and economic benefits that flow from clean electricity as a replacement for traditional fuels, which are at present not incorporated into social and financial decision-making, especially when it comes to pricing electricity.

All of these problems can be countered by critiques from civil society. However, most challenging is the lack of synthesis between the three major citizens' networks that have challenged government policy and corporate practices: environmentalists, community groups and trade unions. Our work at the CCS aims to identify the numerous contradictions within both South African and global energy sector policies and practices and to help to synthesise the emerging critiques and modes of resistance within progressive civil society. Only from this process of praxis, can durable knowledge be generated.

Notes

1. Some of the material in this chapter is revised and updated from a chapter co-authored with Stephen Greenberg and Maj Fiil-Flynn in Bond 2002.
2. Selva's report continued:

   And now, there is a new threat – the dumping of electronic waste, or e-waste: unwanted mobile phones, computers and printers, which contain cadmium, lead, mercury and other poisons. More than 20 million computers become obsolete in America alone each year. The UK generates almost 2 million tonnes of electronic waste. Disposing of this in America and Europe costs money, so many companies sell it to middle merchants, who promise the computers can be reused in Africa, China and India. Each month about 500 container loads, containing about 400 000 unwanted computers, arrive in Nigeria to be processed. But 75 per cent of units shipped to Nigeria cannot be resold. So they sit on landfills, and children scrabble barefoot, looking for scraps of copper wire or nails. And every so often, the plastics are burnt, sending fumes up into the air . . . Inspections of 18 European ports in 2005 found that 47 per cent of all waste destined for export was in fact illegal. In 1993, there were two million tons of waste crossing the globe. By 2001, it had risen to 8.5 million. It is illegal to ship hazardous waste out of Europe, but old electronic items can be sent to developing countries for 'recycling'. (2006)
3. Because Purchasing Power Parity estimates by the IEA are dubious (for example, Zimbabwe's GDP in US$32.7 billion), the actual GDP figures are used. However, South Africa's GDP is far less than US$164 billion, so the ratios indicating South Africa's high carbon/GDP emissions are actually quite conservative.
4. For a full critique of Coega, especially Erwin's role, see Chapter Two in Bond 2002. As of September 2006, there were still plans to offer vast subsidies, very cheap (and reliable) electricity and other investment incentives to the Canadian firm Alcan to pursue a massive smelter at Coega. By 2008, however, the investment appeared to be cancelled or at minimum postponed for many years, due to national electricity shortages.
5. Another reason for low consumption is that people may not be able to afford the cost of appliances required to increase electricity use. A suggestion that has had some support from electricity suppliers is the provision of a 'starter pack' when households are connected, providing a hot plate or a kettle for free (see Leslie 2000: 69). However, the Johannesburg council never followed upon such proposals.
6. SABC News, 1 November 2004. Mlambo-Ngcuka partly blamed the 'universal' entitlement, which meant that in some cases, all municipal residents received their first block free. Yet this was not only good public policy in view of the consistent failure of means tests, but conforms to her political party's 2000 campaign promise: 'ANC-led local government will provide all residents with a free basic amount of water, electricity and other municipal services, so as to help the poor. Those who use more than the basic amounts will pay for the extra they use.'
7. http://www.irinnews.org/report.asp?ReportID=39413&SelectRegion=Southern_Africa&SelectCountry=south%20africa.
8. Reaching the same conclusion, various mid- and late 1990s studies are reviewed in Beall, Crankshaw and Parnell 2002 and also in White, Crankshaw, Mafokoane and Meintjes 1998.
9. The government initially contested these figures as wild exaggerations, but by mid-2004 leading water official Mike Muller admitted in the *Mail & Guardian* (24 June) that in fact, according to a new government survey, 275 000 households were disconnected during 2003, which equates to 1.5 million people – so the Municipal Services Project estimates were 50 per cent 'wrong', but in a way that was too generous to government.
10. The website http://www.polity.org.za/pol/opinion/?show=82281 offers a detailed deconstruction of Erwin's 28 February remarks, concluding: 'In all we calculate that Erwin used the word sabotage three times during a press briefing that lasted 48 minutes'.
11. The World Commission on Dams found that in many cases, the greenhouse gas emissions rate per unit of electricity from large dams exceeds that of conventional energy generation. See http://www.irn.org for more information.
12. Documentation available in the following editions of *Business Day*: 9 March 2000, 26 April 2000, 10 August 2000, 12 September 2000, 17 October 2000, 10 November 2000, 29 November 2000, 25 January 2001, 27 March 2001, 8 November 2001, 12 March 2002 and 16 July 2002.

2

Interrogating Nuclear and Renewable Energy

Muna Lakhani and Vanessa Black

The way in which people use energy throughout the world is causing a lot of harm. This chapter takes evidence from South Africa, focusing on nuclear and renewable energy, but it must be noted at the outset, that we are poisoning our air with the petrol that we use. In countries such as Lesotho, Namibia and Mozambique, we are moving thousands of people to build big dams so that we can create electricity from the flow of water over the dam wall. When we burn dirty coal in our electricity plants, factories and homes, we warm up our planet and cause floods and drought somewhere else. Burning coal also causes many health problems for people using coal at home, or living near coal-fired power stations, especially problems with their lungs and throats. Their houses are also difficult to keep clean and even their washing gets made dirty again. Our coal-fired power stations are among the dirtiest in the world.

The way we are currently using energy means that we will pay more and more for that energy, as the fixed resources (coal, oil and wood) run out. This will mean no petroleum for our grandchildren and those that follow after them and it impacts badly on our health and our enjoyment of life. It also makes it more difficult for us to afford safe and clean energy.

We have to ask ourselves: What are the social, environmental and financial costs of our use of energy? What can we do with energy supply and can we use sources that are better? What are the challenges facing the government and the people in finding a better energy mix? At the same time, we must improve living conditions in informal settlements, hostels and townships. We must find solutions that also allow future generations to live a better life. Quality of life, a healthy environment and sound development are all closely linked. Bringing energy and environmental concerns into all stages of housing delivery and the upgrading process can save money and improve quality of life, not only for us today, but also for generations to come.

The new nuclear threat

Given the growing concern about the inevitable impact of climate change, the nuclear industry is trying to revitalise nuclear power by claiming that it does not release carbon dioxide (CO_2). While CO_2 is not produced at the power plant, large amounts of CO_2 are generated through mining, transport and especially uranium enrichment. Nuclear power generation thus creates over 8 tonnes per gigawatt hour of power that is delivered – much more than renewable energy sources.

Nuclear energy is the energy stored in the smallest piece of matter: the nucleus, or centre, of an atom. When the nucleus of one atom, uranium, is broken (in a process called fission), it forms two new atoms and lets out a large amount of energy in the form of heat. This heat is used to drive a turbine, which then generates electricity. We have one nuclear power station in South Africa at Koeberg, 28 kilometres from the Cape Town city centre.

Nuclear power is not safe. When the nucleus of the uranium atom splits, it also creates new atoms (such as strontium and caesium), which are very dangerous because they are radioactive. This means that these new atoms are always giving off little amounts of radiation. When they go in through people's mouths and noses and find their way into their bones and organs, they can break down cells in those organs and bones. This causes cancer and birth defects.[1] The National Union of Mineworkers says that many people have died from working in nuclear power plants and uranium mines. Some doctors around the world say that communities living near nuclear power stations also die more often from cancer and give birth to damaged children. Nuclear power stations can also have major accidents, such as the Chernobyl accident in the former Soviet Union.

Certain types of radiation can also travel through a person, just like X-rays do, which also causes much harm. It must be remembered that radiation cannot be seen, heard, touched, smelt, tasted or destroyed. Nuclear power also produces dangerous radioactive waste at every stage of the nuclear fuel cycle: from uranium mining, to reactors, to the reprocessing of irradiated nuclear fuel. No one has found a proper solution to the long-term storage of this used fuel and other highly radioactive waste, including depleted uranium ammunition from international nuclear weapon programmes. When Koeberg comes to the end of its life, it will also be contaminated and the whole building will have to be treated as radioactive waste, which will remain dangerous for thousands of years.

All over the world, people are saying 'No nukes!' The nuclear industry in the developed world, particularly Western Europe and the United States, is on its last legs. Germany has put an end to its programme; there are no new orders coming from the United States; France has stopped its new reactor programme and the World Bank has made a decision not to finance any new nuclear power plants. Nuclear reactors can also not be used to minimise greenhouse gases under the Kyoto Protocol. So why is South Africa carrying on with nuclear power? The government's 'White Paper on Energy Policy' says that no decision on nuclear power stations will be taken before all energy issues have been discussed with all stakeholders. But Eskom is carrying on with a new nuclear power programme: the pebble bed modular reactor (PBMR) programme, to be sited at Koeberg or Pelindaba. The reason for choosing Koeberg or Pelindaba is because nuclear sites already exist at these places, making it cheaper for Eskom.

The Nuclear Energy Corporation of South Africa (NECSA) wants to make radioactive fuel at a new factory at Pelindaba, which will also release radiation during the manufacture of fuel. There will be up to nine trucks every day on our roads carrying radioactive products for the next 40 years if full production is achieved. Even just one accident will cost many billions of rand to clean up and the radiation will remain harmful for thousands of years. NECSA has not proven that it can even make the fuel, nor if it can make it safely. A nuclear container designed recently was rejected by the United States and casual labour was used with people working with radioactive material without protective equipment.

There is no debate that radiation kills, maims, causes mutations, is cumulative, causes leukaemia (mainly in children), cancers, respiratory illnesses and attacks the immune system – with children, pregnant women and the elderly being the most vulnerable. The only disagreement is about what is considered an allowable dose. However, there is no such thing as a safe dose of radiation. The only people who say that radiation is safe are those who make money from radioactive processes and they cannot be trusted, as they have proven so far, here and overseas.

Earthlife Africa is alarmed at the potential for dangerous nuclear accidents should Eskom continue with its plans for the PBMR programme. The fuel for the new PBMR is a graphite-wrapped pebble, the size of a tennis ball, containing enriched uranium kernels. The plant to manufacture these fuel balls will be built at Pelindaba, near Tshwane (formerly Pretoria), but the raw material – enriched uranium – will need be imported via a harbour in KwaZulu-Natal, probably Durban.

Eskom officials plan to build approximately 216 PBMRs. Of these, 24 reactors are earmarked for local energy generation and Eskom hopes to sell the remainder on the international market. This means that in addition to transporting enriched uranium, the manufactured fuel pebbles will need to be transported from Pelindaba to various locations as far as Cape Town, as well as to Durban for export.

Radioactive waste

A dangerous feature of the present nuclear policy is its treatment of waste. A metal smelter melts metal into blocks, called ingots. This in itself is not usually a problem, but at Pelindaba,

NECSA wants to build a smelter for melting radioactive waste that has piled up over many years. The corporation says it needs to melt the metal to stop people from getting hold of the technology that makes nuclear weapons.

However, internationally, the waste is treated by cutting it into small pieces (which has to be done for melting anyway), crushing it, so that it cannot be used again and covering it with ceramics (clay), so that it is in a hard covering, which will also not allow people to recover the metal pieces to use as weapons.

All radioactive waste should be stored above the ground, so that it can be monitored, especially so that the radiation can be measured, to see if there are any leaks, which can then be fixed quickly. The melted metal from Pelindaba, which will still contain radiation, will be sold as scrap, releasing the danger to the public, so that our spoons or toasters can be made from it, bringing the danger to our homes.

If we allow the smelter to go ahead, it will be very dangerous for the workers, as they will be working in a room with no special filters; even the special filters on the smelter will let radiation escape. Over 5 kilograms of radioactive dust will escape every year and be spread by wind year after year, increasing the harm it can do to all life in the area.

In the United States it is normal for the radius around which such a smelter has an impact to be 160 kilometres. In South Africa, people live next door to Pelindaba. NECSA has also said that it wants to 'commercialise' the smelter when it has finished melting the waste at Pelindaba. This means that our energy minister and the energy minister of another country can agree to allow radioactive waste to be imported to South Africa, to be melted here, releasing more radiation around us and selling that scrap metal again. Already, radioactive scrap is being dumped in Somalia, Eritrea, Sudan and Mozambique – we cannot be next.

Nuclear fuel in motion

At the height of proposed PBMR production, 9 trucks would drive through Durban every week, carrying enriched uranium to Pelindaba and another 31 trucks would return carrying fuel pebbles for export. A total of 231 012 000 kilograms of radioactive materials would be travelling on our roads during the 40-year lifetime of the reactors – an average of 9 trucks a day, as well as 145 trucks carrying chemicals. (These figures are based on information provided by Eskom in its environmental impact assessment [EIA] process.) Durban residents should be concerned about the enriched uranium transported through the harbour up to Pelindaba. Even if only the first demonstration PBMR is built, 1 000 kilograms of enriched uranium will be transported annually.

Eskom claims it will be transported in specially constructed canisters, but the information provided in the corporation's EIA says that the canisters are only designed to withstand a drop of 9 metres. So if a truck had an accident on a bridge and the canisters fell, they could easily split open and expel the enriched uranium, described by Eskom as a fine powder. At wind speeds of only 3 metres per second, this can blow 10 kilometres within an hour, and Durban's mean daily wind speeds are at least 4 metres per second. In this instance it is very likely that the enriched uranium will not only contaminate land and water sources, but will also be breathed in by many people.

Earthlife Africa is concerned that emergency services in Durban would provide an inadequate response in the event of a nuclear transport accident, given recent statements made by emergency personnel and the response to transport accidents such as the recent asbestos spill on the M7. Furthermore it would be difficult to evacuate such a wide area in an emergency. Hence over the last few years, insurance companies have been amending household insurance policies to specifically exclude any damage or loss caused by nuclear fuel, waste and its associated radiation from their cover.[2]

Used fuel poses the greatest risk in the nuclear chain. Used fuel will need to be transported from the reactors at some point and this waste then needs to be safely stored. There is the possibility that other countries that buy reactors and fuel from South Africa, will demand that the highly radioactive used fuel be returned to us as the country of origin.

Transport accidents involving used fuel could be a calamity. Graphite, the casing of the fuel pebbles, burns readily in air if exposed to temperatures of greater than 800 degrees Celsius. Diesel burns at 1 010 degrees Celsius and magnesium alloy wheels burn at far higher temperatures. Thus in an accident where there is a fire, it is reasonable to assume that the graphite balls will combust. Water cannot be used to put out a graphite fire. Used pebbles are also prone to mechanical damage such as chipping, cracking and the separation between the outer graphite and silicon dioxide layers exposing the uranium/graphite to fires, etc.

In short, South Africa has no policy on how to manage radioactive waste. Medium- and low-level waste is transported to Vaalputs in Namaqualand for long-term disposal. Highly radioactive used fuel from the existing Koeberg reactors has not been moved here because of the great danger. Instead the government spent R80 million on new high-density storage racks.

A bad financial deal

Nuclear energy is not cheap. But Eskom continues to spread false promises of delivering cheap power to the poor. The economic viability of the PBMR is based on spreading the start-up costs across 216 reactors, some of which need to be sold on the international market, which has shown little interest. However, the government and Eskom continue to withhold crucial information on the financial feasibility of the PBMR from public scrutiny.

Ten years ago Eskom was claiming that the first demonstration reactor would cost about R1 billion and would create thousands of jobs. But Eskom now concedes that R1.5 billion has been spent on the design and feasibility process to date. An additional R500 million was authorised by the government in November 2004. Another R12.9 billion is needed to build the first pilot reactor and the fuel plant. These costs exclude decommissioning of the facilities – estimated at a minimum of R100 million, fuel costs for 40 years, ongoing (high) cost of maintenance and storage of the radioactive waste for at least 240 000 years. The existing Koeberg reactors were built at a cost of three times more than originally promised.

Nuclear energy generation costs in the United Kingdom turned out to be double what the government had originally claimed. The last reactor built in the United Kingdom in 1995 cost US$3 000 per kilowatt of capacity – nearly ten times more than it costs to build a gas-fired power plant. British taxpayers are presently faced with a £48 billion bill for cleaning up historical nuclear contamination. Even the World Bank is sceptical: 'Nuclear plants are thus uneconomic because at present and projected costs they are unlikely to be the least-cost alternative. There is also evidence that the cost figures usually cited by suppliers are substantially underestimated and often fail to take adequately into account waste disposal, decommissioning and other environmental costs.' (World Bank 1992)

There is a limited export market for PBMRs. Eskom has been trying to sell its reactor internationally for at least eight years, with expensive international trips and marketing campaigns, but has not been able to show written evidence of any international interest in purchasing PBMR units. Most countries in Europe are phasing out nuclear power, with the exception of France, which has its own nuclear industry. French nuclear company, Areva declared that the PBMR was 'not competitive to generate large-scale electricity'. The United States also has its own nuclear industry, which withdrew support for the PBMR in 2002. The most likely market is East Asia, but China is developing its own reactor models (including a pebble bed).

Not 'Proudly South African'

While Eskom claims the PBMR is a 'Proudly South African' product, this design was actually purchased from Siemens after Germany abandoned its programme. The German programme was shut down because malfunctioning fuel balls became stuck, because of dislocated graphite tiles on the inside of the reactor and because of a large accident at Hamm-Uentropp in May of 1986, which the regional authorities tried to pass off as 'fallout from Chernobyl'. Not only are we trying to tweak unproven, hand-me-down technology, but we will also be importing many of the key parts.

The PBMR also creates few jobs. Most components of the demonstration unit will be imported from foreign companies because we do not have the capability to produce these highly specialised items. During 2005 the following contracts were awarded: Mitsubishi Heavy Industries – contract to build the turbine machinery at a cost of US$12 million and Uhde, a division of Germany's Thyssenkrupp Engineering – contract to build the fuel plant and infrastructure at a cost of US$20 million. This might explain why the thousands of jobs promised by the PBMR programme won't be delivered: only 80 full-time jobs will be provided, mostly for highly skilled people. At present, the PBMR company staff (about 50 full-time employees) earn an average of R480 000 per year, or R40 000 per month.

As another example of *not* proudly South African values, the PBMR appears rife with vested interests. Cabinet appointed Maurice Magugumela as the chief executive officer (CEO) of the National Nuclear Regulator (NNR), the government body tasked with safeguarding the public from nuclear harm. The NNR also gives the final go-ahead to a reactor by issuing the operating licence. Magugumela was formerly the safety and licensing manager of the PBMR company for five years.

Reuel Khoza, the former chairperson of Eskom had a 28 per cent interest in IST Holdings through his own black economic empowerment (BEE) companies when IST was awarded a contract of R260 million for development of the PBMR by Eskom. Louisa Zondo, the previous CEO of the NNR, was on the board of a company that owns a 25 per cent stake in IST.

Safe, clean and simple energy options exist

We all agree that fossil fuels cause environmental and health problems and we need to move away from these unsustainable fuels. The question is through what means? Energy conservation is the first step. South Africans (mostly at household level) reduced energy consumption in 2004 by 198 megawatts (= 1.2 PBMRs), 30 per cent more than the government's annual target.

Public funds should be used to ensure safe, clean and reliable energy delivery to all South Africans. We should invest in those technologies that are at the cutting edge of energy provision: technologies that are sustainable, proven, decentralise power generation to where it is needed and that optimise job creation. Twelve billion rand can buy:

- one 165 megawatt PBMR = 80 full-time jobs + 1 400 construction jobs for one year; or
- 1 700 megawatts of wind power = 850 full-time jobs + 3 000 construction jobs, which can be supplied locally; or
- 5 700 megawatts of solar PV (photovoltaic) = 680 full-time jobs + 8 800 construction jobs, which can be supplied locally; or
- 795 megawatts of generating power saved by providing solar water heating for 1.2 million houses, thereby improving the quality of life of six million people.

Well-proven and commercially viable renewable technologies exist, but haven't been tried at scale in South Africa. For example, wind is a tried and tested technology that one can set up really quickly with relatively unskilled labour and it is easy to build turbines locally. There are numerous sites across the country where wind resources are sufficient for energy generation. Wave, tidal and solar thermal power will deliver bulk industrial supplies safely and reliably – waves and tidal currents never stop.

Our wind and solar resources are amongst the best in the world. If the European Union can develop solar thermal technologies, why don't we, with our more reliable supply of sunlight?

Even if we disregard the dangers associated with nuclear power, renewable energies are better for the economy. The international markets for these safe technologies are growing by up to 40 per cent per year, with the nuclear market growing (at best) a few per cent a year.

What can be done instead?

Because of global warming and pollution at local levels, we need to use less oil for transport and less coal for electricity and industry. Research and development money must be spent on alternative, environmentally friendly energy sources such as wind, solar, wave and biomass. There are a number of ways we can better use energy and make energy safely and cleanly.

Energy efficiency

If we use less electricity to do the same jobs we are being energy efficient. Some fridges and stoves, for example, use less electricity than those that we used in the past. Compact fluorescent lights (CFLs), which use less electricity for the same amount of light are now advertised on television under Bonesa. If we put in ceilings and insulation and make sure that our buildings and windows face north, we can also reduce the energy we use for heating our homes and factories. Eskom was able to reduce its electricity consumption at the head office by 34 per cent, by implementing energy efficiency.

Solar water heating

If we were to use energy directly from the sun to heat water in our homes and factories, we would save half of the electricity that we normally use from the national grid. In some countries, governments have passed laws that compel people to use solar heating. If everyone were to use solar water heaters in South Africa, we could do away with one 2 000 megawatt coal-fired power station, or twenty pebble-bed reactors. Solar cooling can also be used instead of normal air-conditioners. Large mirrors reflect the heat of the sun to one spot on the tower, making that spot very hot – this heats water to make steam, which then turns a turbine to make electricity. There are different kinds of mirrors that can be used – those that are quite flat, or they can be curved (like half a pipe) and a pipe for water can run through the middle.

Solar thermal

Solar power can also be used to generate large amounts of electricity, by concentrating the power of the sun with mirrors or lenses, like a giant magnifying glass. This very hot process easily and quickly turns water into steam, which can then drive a turbine – it is exactly the same generating process as coal-fired power stations, except that the source of heat is the sun. If we used the world's deserts for electricity generation, we could supply the whole world with only 2 per cent of the land covered by deserts. We can also use low-temperature solar thermal for the drying of food, for example.

Solar panels

PV – photovoltaic – panels turn sunlight into electricity. The sun can also be changed to electricity through solar panels, which are normally linked to batteries. Because lights, radios and television sets do not use a lot of electricity, they make the best use of solar panels. Solar panels are also great for people who live far away from grid electricity. Since the demand for solar energy is growing all the time, the cost of manufacturing solar panels is coming down every year.

Wave energy

The waves at the edge of the ocean can also generate electricity. Throughout the world, governments and businesses are conducting more research on wave energy. This is the energy held in rising and falling waves at sea, which makes a wave generator go up and down and so make electricity. This is already happening commercially. As the storms in Cape Town and elsewhere have shown, South Africa's coastline has excellent potential for wave power generation and Stellenbosch University has already produced working models. For example, it is estimated that 0.2 per cent of the ocean's wave energy could supply the current worldwide demand for electricity. Every metre of coastline in northern California provides enough energy to power twenty average American households, who use more electricity than South Africans. Rough calculations show that 40 metres of wave-front could produce enough power to run the Point Hotel in Cape Town (200 kilowatts per 40 metres of wave) and with only 1 kilometre of wave, we could generate enough power for Cape Town. Wave energy is captured in a very easy way – the waves go into a pipe at the bottom – this pushes the air in the tube out – while it is doing this, it can turn a fan (turbine), which generates electricity.

When the wave drops, air comes rushing back in, which can turn the turbine again, to make more electricity.

Wind energy

Wind energy is one of the fastest growing industries in the world and the cost of wind power has been coming down every year. It is also competitive with coal and gas. If you take into account the health costs of coal, wind has already been found to be cheaper than coal in the United States. In some countries wind already produces over 10 per cent of people's energy needs and by 2010, wind energy will supply over 10 per cent of Europe's power needs. Some analysts predict that wind could supply up to half of all global energy needs by 2100. Wind is better than coal and nuclear power because it can often create electricity close to where it is used; there are very few impacts on the environment; it creates local jobs and South Africa has great wind resources along the coastline and the escarpment. We need to use this energy source more in the future.

Hydro-energy

For centuries, people have been using the energy from small amounts of moving water to grind grain – like the Josephine Mill, next to the Newlands sports ground in Cape Town. In the last 100 years, however, engineers have built massive dams to hold back large amounts of water and then let it out to run through big turbines, to generate hydroelectricity. But these big dams have equally huge environmental problems (like the Narmada Dam in India) and governments are moving back to small hydroelectric generation (or micro-hydro), as this has less impact on the environment. We can build micro-hydro schemes on small rivers and equally small dams. If we use local technology and skills, we can also create local jobs. In some countries such as Sri Lanka, micro-hydro can supply up to 90 per cent of people's energy needs.

Bioenergy

Bio means 'life', so biomass is the raw material of living things. Biofuels are the kind of fuels we can get from biomass and the outcome is bioenergy. So we can burn biofuels directly, such as wood, but we can also change biomass, such as sugar cane or beet, into gas. We can also change biomass chemically into liquid fuels, such as ethanol, which we can then use to generate electricity, or burn as transport fuel. The leftover mush (or slurry) can then be used as compost. We can call wood and other biofuels sustainable and renewable, if we harvest them in a way that does not destroy the environment. In many countries, small biogas digesters are used to produce gas for homes or communities, but in Denmark, twenty large biogas plants currently digest wastes from animal and food-processing wastes. We can also capture usable gas from sewage.

Geothermal energy

When the heat from the centre of the earth (geothermal energy) is close enough to the surface, we can use it to heat water and so generate electricity. Global usage is growing very fast and now stands above 8 000 megawatts. Geothermal energy is not dependent on the weather and can be utilised 24 hours a day.

Tidal energy

A good answer to those who say that all renewable energy is intermittent is to suggest that – besides wave, wind, geothermal, and micro-hydro (which are generally very predictable) – we use tidal energy. Water in the oceans is constantly moving, at different levels underwater and never stops. These tidal currents are very strong and are responsible for some of our plastic bags being found in Australia. Similar technology for micro-hydro and wave energy can be used here and is already being tested commercially.

Storing energy

Fuel cells are devices that combine the basic elements of hydrogen and oxygen to produce energy and water. Many people see them as a good way to store energy from natural sources, such as solar and wind. This is because fuel cells need some energy first to produce hydrogen, which can then be made into electricity. Fuel-cell technology is growing fast: some of the big motor companies want to have products on the market by 2003. Many cities around the world are already testing fuel-cell engines and these engines could soon replace the noisy, polluting car engine we know so well. Fuel cells produce energy on tap and can also be used as small, portable power plants (much better than pebble-bed reactors). Another advantage of hydrogen fuel cells is that we can use intermittent (as well as other) renewable technologies to produce the hydrogen.

Pumped storage

This is already being used in South Africa, but not using renewable technologies. Pumped storage refers to a process in which (usually) water is pumped up to a higher level, using energy when there is little or low demand and then released to make more energy when required – for example, at times of high demand, like winter mornings and evenings. Renewable energy is very good for this application, as it can be used whenever the resource is available and then released when needed.

Gas

Because natural gas gives off less CO_2 when it burns, scientists see it as a fuel which can bridge the gap between the old days of dirty coal and oil to the coming clean-energy world, based on renewable energy and hydrogen power. In this period of the early 2000s, we are using gas for over one-fifth of our global energy needs. Although gas is also a fossil fuel, it produces much less CO_2 than coal or oil. Southern Africa has abundant supplies of gas.

Notes

1. Although enriched uranium is only mildly radioactive, ingestion through the mouth and nose is extremely harmful because of its chemical toxicity, which is comparable to lead, according to the World Nuclear Association. The main chemical health effect ascribed to uranium in humans has been damage to the kidneys. Recent research shows that ingested particles can enter the bloodstream from the lungs or stomach where they may exert systemic toxic effects. Acute pulmonary effects relating to chemical toxicity have been observed in rabbits. Insoluble uranium particles can remain in the lungs for many years causing chronic radiotoxicity in the alveoli, potentially leading to cancer. The National Radiological Protection Board in Britain conceded in 1995 that 'there is in fact no threshold radiation dose under which one wouldn't risk growing a cancerous tumour – in other words even small doses can make one ill'. Every day, 452 used fuel pebbles will be released from the reactor core and become waste –translating into a daily amount of 226 000 curies of radioactivity generated daily at one reactor, which will remain dangerous for thousands of years. These are enormous quantities. A single curie of iodine 131 – one of many isotopes created through nuclear fission – could make 10 billion quarts of milk unfit for continuous consumption (US guidelines). The PBMR EIR final report clearly shows strontium-90 as a by-product of the fission process in a PBMR reactor. This isotope has been directly related to incidences of leukaemia among children of fourteen years old or younger.
2. The following are a small sample of nuclear transport accidents reported in the United States: The Critical Mass Energy Project (part of Ralph Nader's 'Public Citizen' group) tabulated 122 accidents involving the transport of nuclear material in 1979, including 17 involving radioactive contamination. For example, two canisters containing radioactive materials fell off a truck on New Jersey's Route 17 in September 1980. The driver did not notice the missing cargo until he reached Albany, New York. In 1986, a truck carrying low-level radioactive material swerved to avoid a farm vehicle, went off a bridge on Route 84 in Idaho and dumped part of its cargo in the Snake River. Officials reported the release of radioactivity.

PART 2

SOUTH AFRICA'S CARBON INVESTMENTS

3

Low-Hanging Fruit Always Rots First

Observations from South Africa's Carbon Market

Graham Erion with Larry Lohmann and Trusha Reddy

In the decade since the signing of the Kyoto Protocol in 1997, carbon trading has come to dominate discussion about climate change mitigation, even among many of the countries and organisations that originally opposed the idea. Yet while the European Union is singing a new tune with its own massive trading scheme and organisations such as the Worldwide Fund for Nature (WWF) sponsor conferences teaching corporations how to trade carbon credits, opposition to trading is also growing, both intellectually as well as around actual projects. This chapter will detail some of these struggles through an in-depth analysis of the carbon market in South Africa.

There are many reasons why South Africa is such a suitable candidate for this review. As the only African country with any serious Clean Development Mechanism (CDM) project development, the success or failure of the CDM in South Africa will have enormous implications for the carbon market on the rest of the continent. Moreover, with nearly two dozen projects in various stages of development, South Africa's variety of methodologies and project developers is relatively representational of the global carbon market, even though it has far fewer projects than some other countries. This allows a relatively small sample size of South African projects to represent overall trends in the global carbon market. Finally, South Africa's rich history of social mobilisations, especially during the apartheid era, provides a unique context to study the opportunities for civil society to influence carbon trading projects and policy in a host country.

Before discussing the projects examined in this chapter, it is necessary to provide some context for the development of the carbon market in South Africa. This context will be explored by asking three questions:

- what types of projects are being developed?

- who is developing them? and
- how were the five projects chosen for this study?

As of September 2006, the South African Designated National Authority (DNA) had reviewed twelve different projects. Of these, only two have completed the entire validation process: the Kuyasa low-cost urban housing energy upgrade project in Khayelitsha (outside Cape Town), which was the first project in Africa to receive validation by the CDM executive board and the Lawley Fuel Switch Project in Gauteng. There were three other projects in late stages of validation including the Durban landfill gas projects at the Mariannhill and La Mercy sites, a PetroSA biogas project, and a fuel switching project with SA Breweries. The other seven projects are in earlier stages of the validation cycle. Among these projects are a variety of methodologies being used to reduce emissions including fuel switching, methane capture, small-scale hydro and biogas.[1]

A final point of context is necessary before discussing the project case studies, namely to explain why they were chosen as representative of the South African carbon market. Briefly, the projects studied include Durban Solid Waste's landfill gas capture, Sasol's fuel switching, Bellville's landfill gas capture and Kuyasa's low-income housing energy upgrade. Collectively, these projects represent some of the most popular methodologies, a mix of project developers and geographic diversity. All are at various stages of validation. A fifth project, Climate Care's energy-efficient light bulbs, is also examined for the implications for the voluntary offset market.

It must be noted that these projects are not intended to give a complete picture of all the trends in South Africa's carbon market. With a dozen projects in the validation cycle and more on the way at the time of writing, such an undertaking is outside the scope of this analysis. Yet through these projects some general observations can safely be made about South Africa's carbon market and its ability to further the global struggle against climate change. Each project has fundamental flaws that unveil not only design weaknesses, but also core problems in the way the carbon market relates to real people and ecologies.

Landfill gas capture in Durban

Any serious discussion of the CDM in Africa should begin with the landfill gas capture project in Durban. This was the first CDM project on the African continent and was initially proposed in 2002 when South Africa hosted the World Summit on Sustainable Development. South Africa was promised US$15 million from the World Bank's Prototype Carbon Fund (PCF) in start-up capital, one of the first projects the PCF ever supported. Finally, with the possible exception of the Plantar sinks project in Brazil, this is the most controversial CDM project to date and has easily garnered the most attention of international activists and media (see Carbon Trade Watch 2003; Reddy 2005a; Vedantam 2005).

On the surface, the Durban Solid Waste project is straightforward: at three landfill sites across the city – Bisasar Road, La Mercy and Mariannhill – wells are drilled to capture landfill gas that would otherwise be released into the atmosphere. The gas is mostly carbon dioxide (CO_2) and methane, a greenhouse gas that is 21 times more potent than CO_2, occurring in a variable ratio of 40–60 per cent. Currently landfill gas is captured and flared at the Bisasar Road and Mariannhill landfills, but this is only about 7.4 per cent of the potential gas that could be captured (PCF 2004b: 3). The proposed project plans to substantially increase the efficiency of the gas capture up to 83 per cent by 2012 and averaging 45 per cent collection efficiency over the 21-year life of the project (PCF 2004b: 4). Once the gas has been captured, it will be burned to generate electricity for use by industrial consumers, thus offsetting coal emissions from the electricity these industries would have used normally. Had this project got underway in 2004, it was claimed that it would offset a total of nearly 2 million tons of CO_2 equivalent (CO_2e) by 2010 (PCF 2004b: 26).

This project is claimed to be additional, since it is capturing methane gas well beyond levels proposed by the regulations and the capacity of local officials, plus local industries would not want the electricity without the carbon credits subsidy, since it would be cheaper to get power from coal. Lindsay Strachan, engineer at Durban Solid Waste and one of the main project proponents admits, 'Even if this wasn't a CDM, we'd still have to take out the gas, but not all of the gas, just what's required by the regulations'.[2]

One need not look far to find reasons why opposition to this project has been so fierce. For starters there is the location of the landfill sites: the La Mercy site might be well away from residential areas, but both the Mariannhill and Bisasar Road sites are in residential areas. This is less of a problem at Mariannhill, as there are large buffer zones on all sides of the landfill. In contrast, there is no buffer zone around the Bisasar Road site where the landfill is literally within a few metres of residential houses on two sides and across the street from a school on a third. Worse still, Bisasar Road is the largest landfill site in Africa and one of the largest in the southern hemisphere.

A short history of Bisasar Road

To tell the story of Bisasar Road one must begin not with the landfill, but with the Group Areas Act of 1961, whereby the apartheid government relocated the Indian population across Durban to the area known as Clare Estate, where Bisasar Road is situated. At the time of the resettlement there was an enormous quarry on Bisasar Road that was lined with trees and green space. In 1980 when the local government was running out of landfill space, they converted the quarry into the Bisasar Road dump. The fact that this was almost an entirely Indian neighbourhood during the time of apartheid is not coincidental.[3]

From the very beginning Bisasar Road was a controversial and contested site. Many of the Indians in Clare Estate were relatively middle class and thus had the resources to quickly organise against the dump. The response of the city was to announce that the dump would close in 1987. Seven years later the city reneged on this promise, but assured the community that the dump would close in 1996 and then be converted into a recreational and sporting site (Reddy 2005a: 3).

In 1996, there was further procrastination. The city began a public consultation process ostensibly intended to establish a permit to close the dump (South Africa requires permits not only to open a landfill site, but to close it as well). It was at these meetings that local resident Sajida Khan – who lived directly across the street from the landfill until her death in July 2007 – found out that the permit process was actually intended to extend the life of the dump rather than close it. When Khan discovered this, 'I just went nuts! I wouldn't let anyone else talk. I was just so angry' (cited in Reddy 2005a: 3).

Khan quickly channelled her anger into an organised campaign. With ten public schools within 1 square kilometre of the landfill, Khan chose to target children in her campaign and through this 'the parents and other people would get roped in' (Reddy 2005a: 3). Khan's campaign tactics included placard demonstrations, blockades of the dump (this was the only activity little children were not involved in for fear of injury), a community-wide petition with 6 000 signatures and a media blitz.

Despite Khan's best efforts, the permit to extend the life of the dump was granted. To make matters worse, in a wealthy white-dominated suburb to the north of Durban, the Umhlanga landfill site was rapidly shut down, as it was 'earmarked for up-market property development' (Reddy 2005a: 3). The waste from this site was re-routed to Bisasar Road.

Health effects

In 1996, not only was Bisasar extended and given yet more rubbish from Umhlanga and from another site in Umlazi, Khan also developed cancer and a second bout in 2005–07 proved fatal. From her informal surveys of the neighbourhood, Khan claimed that seven out of ten households in the area of Clare Estate closest to the landfill reported at least one person developing cancer. Among these victims was Khan's own nephew, who died of leukaemia.

For Khan and other residents in Clare Estate there was only one place to lay the blame for their poor health: the dump. Prior to the 1990s there were very few government regulations on waste management and thus Bisasar Road was able to have a medical waste incinerator on its site and accept other forms of hazardous waste (Reddy 2005a: 3). Even when stricter regulations were put in place and the landfill ceased incinerating hazardous waste, Khan still cited unsubstantiated studies where the limits of waste emissions considered potentially hazardous were exceeded in hydrogen chloride by 50 per cent, cadmium by 200 per cent and lead by more than 1 000 per cent. Limits for suspended particulate matter were also exceeded (Reddy 2005a: 5).

It is not surprising that Khan's assessment of the health impacts of Bisasar Road would be disputed by officials at Durban Solid Waste. According to Strachan, 'We've brought in experts to assess the health risks. Their main concern was formaldehyde, but the health experts couldn't discern if it was burning from Kennedy Road or if it was landfill.'[4] Strachan argues that any health threats in the area would come from the informal housing community on Kennedy Road that regularly burns wood and other materials for heating and cooking as there is no electricity. As to Khan's survey of ten households in Clare Estate with high rates of cancer, Strachan questioned her methodology and research qualifications, concluding that her research was 'absolutely codswallop!' Furthermore, Strachan pointed out that there is a one in four cancer incidence rate in Durban and therefore 'how do we know [these people's cancer] is from the dump? With those odds, it could be from anything.'[5]

Whether cancer rates can be attributed to the landfill or not, a growing concern in Clare Estate is that the CDM project will create more air pollution and potentially adverse health effects. Khan calculated that each year, the proposed methane electricity generators will pump out 95 tonnes of nitrogen oxides, 319 tonnes of carbon monoxide, 323 tonnes of hydrocarbons and 43 256 tonnes of CO_2. Nitrogen oxides are respiratory irritants and exacerbate asthma, carbon monoxide reduces the oxygen-carrying capacity of the blood and carcinogens such as benzene and butadiene could be found in hydrocarbons (Reddy 2005a: 8). These figures should only be taken as estimates, however, as the scientific validity of Khan's calculations was not confirmed.

The issue of closure

Although members of the Clare Estate community remain concerned over potential health impacts from the CDM project, their main point of contention with this project is the widely held perception that it will further prolong the life of the landfill site. Strachan adamantly rejects this and argues that the landfill gas must be captured either way, so it doesn't matter if the landfill is still accepting waste or not. In addition, Strachan is just as insistent that 'the dump is closing . . . the city is saying we'll close it'.[6]

But the way the city is going about this is through the creation of a waste transfer station near the south end of the landfill, so that when the site closes waste can be transferred to a new landfill further away. According to Strachan, the environmental impact assessment (EIA) for the transfer station is costing the city about R1 million, which could then be expanded to include closure for the landfill: 'The transfer station is the start of the closure process.' Ironically, Strachan blames the residents' opposition to the transfer station – which they see as just further development and pollution in their neighbourhood – as an impediment to closure: 'If you walk into a room you're just heckled, you can't talk to people. So the dump continues.'[7]

Strachan says he sympathises with local residents and claims to be much more concerned about the viability of the CDM project than the continued operations of Bisasar Road: 'I haven't received a closure demand in two years; they're now driving the anti-CDM train; they should keep driving the site closure train and make it quite clear that if you close the landfill, we want this gas project as long as the landfill is closed.'[8]

There is evidence to contradict Strachan's view that the CDM and continued operation of the landfill are not related. For example, in the 2004 project design document that Strachan helped to prepare, the baseline methodology for this project states:

> All three landfills have remaining capacity and, with the exception of La Mercy, can continue to operate throughout the crediting period. Considering the high costs of developing new landfill sites, it is not reasonable to expect that the municipality would close these landfills before they are full, nor are there any plans for the construction of replacement sites. (PCF 2004b: 8)

The crediting period referred to here was seven years, with two optional renewals of the same amount. Thus when Strachan claims 'the dump is closing', it's possible that he is implying – as the official document does – a 21-year timeline. In addition, a senior engineer at Durban Solid Waste who has worked at the landfill for four years admitted he did not know anything about an impending closure: 'What closure? There's room here for at least another decade of landfill.'[9]

While there is still no irrefutable evidence that the CDM project is what is keeping the Bisasar Road landfill open, there does appear to be a causal link between the two. This link is carbon credits, or to be more precise, an estimated R20 000 per day of potential carbon finance that could be coming into Durban, according to Strachan's calculations. Yet when he was asked whether these calculations involve the landfill site being open or closed, Strachan told a local newspaper reporter, 'The site has the potential to produce 8 000 cubic metres of methane an hour and closure would bring that down to 7 000 cubic metres, so the difference is somewhat negligible' (Robbins 2002). Whether a difference of 12.5 per cent of production is 'negligible' in a US$15 million deal with the World Bank's PCF should be treated as more than a rhetorical question. When asked, Strachan refused to indicate whether he used the higher or lower number in his discussions with the Bank and in the project documentation.

One final issue to mention in terms of the closure is the informal housing settlement on Kennedy Road, directly adjacent to the landfill. As some of the apartheid laws began to relax in the late 1980s, in particular the Group Areas Act, a sizeable group of low-income African people moved into the area around Kennedy Road that runs along the western border of the Bisasar Road landfill. This settlement illustrates the unique tendency for groups of people to gravitate *towards* waste management facilities where waste-picking and other scavenging opportunities offer an alternative means of survival when government resources are limited and unemployment rates are extremely high (Horton n.d.).

This intentional settlement next to the landfill creates obvious conflicts with the rest of Clare Estate who had the landfill involuntarily imposed on them. An employee of Durban Solid Waste describes the divergence between Clare Estate and Kennedy Road: 'One community built up *because* of the landfill, while the other wants the landfill closed' (Horton n.d.: 99).

In the struggle around landfill closure and the CDM, the strategic support of the Kennedy Road community by Durban Solid Waste is considered a very high priority. To this end, the World Bank commissioned a formal recognition of the Kennedy Road community, which Raj Patel observes, 'seems central to the community's support of the project . . . in contrast with richer activists [who ignore Kennedy Road]' (Horton n.d.: 99).

Moreover, in eliciting the support of the community for the CDM project, Strachan reportedly offered 45 jobs and three bursaries to children from 'affected communities to study engineering, possibly in Uganda', although it should be noted that within the Kennedy Road settlement, this figure is believed to be 50 scholarships (Horton n.d.: 99). The reality is that the project will offer six highly skilled jobs and three bursaries over its proposed 21-year lifespan, with no guarantee that Kennedy Road residents will benefit.

Whatever the figure, community leaders in Kennedy Road are convinced that the continued operation of the landfill offers them the best opportunities for economic advancement, while they also remain in relative proximity to the city centre. This resulted in active opposition to the campaigns by Khan and others in the Clare Estate to close the dump and caused a general breakdown of interracial, interclass community relations. For her part, Khan pointed the finger at Strachan for using this divide-and-conquer strategy and claimed, 'I have nothing against these people . . . I am fighting for all of us, no one wants to live next to a smelly dump'.[10]

The present status of the project

In June 2002, just after the PCF signed an emissions reductions purchase agreement with Durban Solid Waste for the CDM project, Khan filed a lawsuit against the eThekwini municipality and the Department of Environmental Affairs and Tourism for negligence in permitting the continued operation of the Bisasar Road landfill. After three years of delays, the case was due to be heard in October 2005.

However, Khan's health had rapidly deteriorated due to another bout of cancer in late 2005 and the case was postponed, as was another court date a year later in September 2006, for the same reason. In addition to the legal action, an appeal to the minister of the Department of Water and Forestry at the provincial level delayed project approval, until it was exhausted in July 2006. Strachan estimates the cost of these delays for eThekwini was at least R40 000.[11]

With pressure building from the World Bank for some progress on the project, Strachan went ahead with project development documentation and applications for the two projects at La Mercy and Mariannhill. He did not mention anything in these documents about Bisasar Road. This was a significant concession, as the two smaller projects will provide a mere 3 megawatts of power between them and only 50 000 tonnes of CO_2e emissions reductions, compared to 10 megawatts of power at Bisasar Road and 3.1 million tonnes of CO_2e.

For the time being it seems that Khan's many years of tireless campaigning won a temporary victory in delaying this CDM project. Bank staff appear skittish about the project, indicating that the costs of bad publicity and continual legal skirmishing offset the benefits they see from a US$15 million investment.

While Khan's battle over the Bank's stake at Bisasar and hence the possibility of it being declared a CDM, was apparently won, the war over closure appears lost. There have been no recent decisions or announcements relating to the eventual closure of the Bisasar Road landfill site and there is little evidence that Durban Solid Waste will bow to pressure to close the dump and remove the waste to another site. If anything, it is likely that with zero-waste measures such as composting, recycling and rubble re-use under consideration as part of Durban Solid Waste's waste management system, the Bisasar dump and future transfer station will remain in use in the residential area for many years to come.

The story of Bisasar Road is in many ways a story of the failure of the carbon market to provide any real options for community development. At no time did anyone give much consideration as to how the project or at least the revenue from the project could really improve the lives of people living next to it everyday (outside of the disputed scholarships). Nor was there ever much thought given to an alternative to extracting methane and burning it to generate electricity that would replace electricity that otherwise would have been generated by burning coal.

Of course, in reality there are many alternatives as to how this project could have been of much greater benefit to the local community. But the carbon credit market demands that there be only *one* alternative. If there's more than one alternative, then there is more than one number corresponding to the carbon saved and you won't be able to assign a single number to the number of carbon credits your project is producing. Other genuine alternatives are classified as implausible: using the money to close the dump down and treat some of the waste; pumping the landfill gas into the nearby Petronet gas pipeline network, so that it would not need to be burned on site; finding ways of using electricity more efficiently or developing more non-fossil community-level power sources. But none of these can be acknowledged as alternatives, because to do so would make it impossible to calculate the credits for the project under consideration. This clearly illustrates one of the ways in which a seemingly technical accounting system can limit the political choices a society can make to small, incremental variations on business as usual.

Sasol's pipeline

From its less than humble origins of supplying embargoed apartheid South Africa with liquid fuels from coal, Sasol has grown to be one of continent's largest companies, with R102 billion in total assets as of June 2006. The company's shares are now traded on the New York Stock Exchange (Sasol 2006).

Sasol's entry into the carbon market was initiated following its decision to build a pipeline to carry natural gas from the Temane and Pande fields in Mozambique. This particular CDM project is unique for a number of reasons. For starters, with an estimated annual reduction of 6.5 million tonnes of CO_2e, it represents one of the largest CDM projects in Africa to date. More importantly, this project has raised some of the most critical questions about the additionality criteria of CDM projects. Through Sasol's inability to persuade its critics, it may also represent the greatest victory to date among those opposing unjust CDM projects in South Africa.

The root of Sasol's additionality issue is its upfront admission in its Project Identification Note (PIN) that its coal mine in Sasolburg was exhausted by 2001 (Sasol 2005: 4). This was a well-known fact at the time, since the drop of production at the mine from 70 million tonnes per year to 2 million had forced enormous layoffs and attracted media attention. Following this, Sasol began trucking approximately 12 500 tonnes of coal per day into Sasolburg from Secunda, a procedure the organisation admitted 'was not an economically sustainable mode of operation' (Sasol 2005: 5).

Hence the company devised two potential options: build a new mine further outside Sasolburg at Vaal, or build a natural gas pipeline to Mozambique. In its PIN, Sasol argues that its baseline scenario was indeed to build the coal mine and thus if it were not for the expected financial grant (through the CDM), they would have chosen coal over the gas pipeline. This hypothetical scenario ignores a number of realities, the first being the fact that Sasol had already lost a court case that prevented them from strip-mining coal at the Vaal site, due to objections from local residents. There was also a problem of time, as the pipeline had already been constructed by the time the CDM application was even contemplated. Many actors in the South African carbon market were suspicious of Sasol's intentions, but no one had uncovered hard evidence to dispute Sasol's claims that the project was additional.

An admission of as much was offered to Graham Erion during a meeting of the South African National Energy Association at the Siemens headquarters in Sandton, outside Johannesburg. At this meeting of energy representatives and lobbyists, Sasol's natural gas supply manager, Peter Geef, gave a presentation on the pipeline and the reasons that Sasol built it. While Geef went through his presentation, including slides such as 'What was this project about?' and 'What made the project possible?' no mention was made of carbon finance (Geef 2005). Following the presentation, Geef was asked whether there were any outstanding financial inputs for this project, to which he responded in the negative. As a follow-up, he was asked if Sasol was indeed pursuing carbon credits to cover the US$1.2 billion cost of its pipeline, whereupon he admitted:

> Yes we are indeed trying to get some carbon finance for this pipeline . . . (But) we have this problem of additionality; we think there's a case to be made for that, we're in discussion with the South African government now and we're trying to make the case for it . . . *The biggest issue is additionality; we would have done this project anyway.* (Geef 2005, emphasis added)

To follow up on Geef's admission, Sasol's greenhouse gas abatement officer, Gerrit Kornelius was contacted in connection with this research. In response to questions about finance, Kornelius pointed to an article from the June 2004 edition of *Global Energy Review* on the project. Although this article goes into great detail about the project's 'financing strategy' and includes a 'summary of financing package', it never once mentions carbon trade financing (Fyfe 2004: 46).

When pressed for an explanation, Kornelius justified Sasol's pursuit of carbon finance on the basis that 'a recent review has indicated that the IRR (internal rate of return) is [at this stage] somewhat lower than envisaged in the original board submission for project approval, and that did not meet the normal hurdle rates for projects – this is one of the arguments for the additionality claim'.[12] Thus Sasol's apparent interpretation of additionality is not in comparison with *what you would have done anyway*, but rather an additional bonus for something *you already did, yet wished were more profitable*.

Unfortunately for Sasol, this interpretation of additionality did not sit well with civil society actors in this arena. Geef's frank comments were distributed across the large network of the South African Climate Action Network (SACAN) and dozens of people learned of this deception. Following a number of failed attempts to hold a meeting with SACAN to convince them to drop their opposition, Sasol appears to have decided to abandon plans for CDM credits on this project (although other projects are being pursued by the company). No further documentation has been submitted to the DNA in the recent past and Kornelius has left Sasol and doesn't appear to have been replaced.

However, there are rumours that Sasol is still hoping to get credits on its project through supplying customers with a cleaner fuel. Large industries with sizeable fuel inputs can thus claim to be reducing emissions, for which Sasol could get credits. Of course, the additionality issue remains the same for downstream sales as it did for upstream production, so the legitimacy of such a project remains in question. Whether or not civil society will react to such efforts and force a similar outcome is another question.

What may be most fascinating about this story is that both sides of the CDM debate can claim it as fodder to support their positions. Those favouring carbon trading (including some in SACAN leadership) see it as evidence that the market functions effectively, since a project that did not meet the required criteria failed to get approval. Yet opponents can also point to some of the more egregious statements made by those involved in this project as the type of thinking and logic that pervades this market. The question is whether Sasol's take on additionality is the exception or the rule. A deeper analysis of future projects in this market will be required, especially anything that Sasol may try to develop in the future.

Landfill gas capture in Bellville

The Bellville project in Cape Town is quite similar to others in its design: drilling wells to capture landfill gas through active extraction, aimed at optimising gas production that would result in a 'conservative' 70 per cent of the gas being captured and utilised (SSN n.d. 'Bellville South Landfill'). Since Bellville is smaller than Bisasar Road, the expected annual emissions reductions from the gas capture and offset coal emissions are only 1.2 million tonnes of CO_2e.[13] As of July 2005, the Bellville project had completed some technical and financial feasibility studies, as well as the preparation of project documentation, although this had not yet been submitted to the DNA.

Comparing the projects in Cape Town and Durban, an initial observation is that the former is being developed under the close supervision of SouthSouthNorth (SSN), a non-profit consultancy that has much more community support than the World Bank's PCF in Durban.[14] There are also some notable differences in terms of the host municipality. In Durban, the environmental planning department includes eleven people and CDM projects are almost entirely handled by Deborah Roberts, who admits 'climate change is something we get to between half past two in the morning and three'.[15] In Cape Town, 106 people work in environmental planning and climate change has its own office, headed by Craig Haskins. Cape Town is also very active in the Cities for Climate Protection programme and boasts an unparalleled expertise on the issue compared with almost any other level of government in the country.

Despite these differences, there are some important similarities between the two landfills, mainly their location in urban areas and the ongoing struggles over closure. According to SSN project developers, 'For the CDM project to happen, the landfill has to be capped. Even with an extension to 2009, the portion that stays open will be capped soon and the portion for 2006 will be capped now.'[16]

How can it be that the landfill must be capped for Cape Town to extract the gas, yet Durban can keep a dump open for 21 years and get 3 millions tonnes of CO_2e per year? This remains a mystery. What is known, however, is that local residents oppose the dump's continued operation. The CDM project's potential need for ongoing dump operation raises questions about whether gas capture projects contribute to the well-being of the local communities in Cape Town, Durban, or anywhere else. The fact that this particular project is looking for certification as a gold standard for the highest levels of environmental and social sustainability makes these questions all the more pressing.[17]

The gold standard

As laid out in article 12 of the Kyoto Protocol, host countries for CDM projects are able to set their own sustainable development criteria for their projects and judge them accordingly on such merits. This has been controversial, as some see the potential of a race to the bottom and a lowering of standards to attract more carbon capital.

Attempting to prevent this situation, SSN with the support of the Climate Action Network (CAN) established a set of universal sustainable development benchmarks in 1999. These efforts were ignored when the market rules were decided. Consequently the carbon market developed in ways the environmental non-governmental organisations (ENGOs) were hoping to avoid, namely widespread 'failure to demonstrate "additionality" and deliver added environmental and social benefits' (BASE n.d.: 1). In May 2003, the WWF released the gold standard code of best practices and criteria 'necessary to deliver real contributions to sustainable development in host countries plus long term benefits to the climate' (BASE n.d.: 3).

The gold standard, which shares strong similarities with SSN matrix, differs from regular CDM project benchmarks in three important ways. Firstly, there are fewer methodologies that qualify for a gold standard and they must fall into the two broad categories of renewable energy and energy efficiency. Secondly, the additionality requirements are ostensibly stricter than the CDM since project developers must show that carbon credits enable the project activity to overcome at least one barrier from a list of five categories: financial, political, institutional, technological and economic. Most importantly, the gold standard seeks to ensure that the sustainable development aspects of CDM project activities are 'maximised' through the obligatory use of 'sustainability matrix Environmental Impact Assessment (EIA) procedures'. These enhanced EIA procedures stress public consultation and evidence that the project contributes to sustainability in three main areas: (a) local/regional/global environment: impacts on air/water/soil quality and biodiversity, (b) social sustainability: impacts on poverty alleviation, access to energy services and human capacity, i.e. empowerment, education, gender and (c) economic development: employment, balance of payments, technological self-reliance (BASE n.d.: 6).

While the Bellville project has yet to receive official registration as a gold standard project by a United Nations Framework Convention on Climate Change (UNFCCC) sanctioned Designated Operational Entity (DOE), SSN already claims the project to be 'in compliance to the Gold Standard Label'. More specifically, the organisation claims the project has 'a positive scoring for all the pillars, with significant contribution in terms of the local, regional and global environment and has scored lesser, but by no means insignificant, contribution toward social sustainability and economic and technological development' (SSN n.d. 'Bellville South Landfill'). These claims will now be critically considered.

To begin with the issue of economic development, SSN admits that the economic development impacts of this project 'would be less significant, this is however counter-balanced by the cost effectiveness of the project due to the potential income from carbon finance and the sale of gas' (SSN n.d. 'Bellville South Landfill'). Thus the impression is that as long as the project is capable of making a lot of money, it can in theory contribute to economic development, depending on how the money is spent.

Yet within the city of Cape Town there is no consensus for how carbon finance from Bellville would be used. SSN hopes to apply the carbon profits from Bellville to other CDM projects in the area that are much less economically viable, such as the Kuyasa energy upgrade (discussed in the next section). Craig Haskins of the city of Cape Town confirmed that discussion on how the revenue will be spent has taken place, but no decision was made in his department, as they did not have the institutional mandate to do so. Should SSN's proposal be adopted by the city council, it is still unclear how taking carbon finance out of the local community in Bellville would further economic development there.

As to social indicators, it seems ironic that a project that is widely opposed by the local community could register a 'by no means insignificant contribution towards local sustainability'. One way to square this circle would be for electricity generated from the landfill to be distributed at low or no cost to the surrounding community, yet such a proposal has not been given any serious consideration in Bellville or Durban.

Finally, turning to environmental sustainability, it seems to be common sense that a project that reduces harmful greenhouse gas emissions would by its very nature deserve recognition as furthering local and global sustainability. However, there is still the very real issue of the lack of sustainable waste management policies associated with landfill gas projects.

Walter Loots, head of Solid Waste for the Cape Town municipality, admits that the present landfill practices are not sustainable, especially in light of lack of available space for landfills. For Loots, the 'real solution to the problem is in sorting and treating waste'.[18] Approximately half of the waste in Cape Town landfills comprises of biodegradable organic material. If this was separated out from the non-organic material, the city of Cape Town would be able to vastly decrease its need for landfill space, as well as capture a much higher amount of methane.

Methane is generated from rotting organic material, yet when this is mixed in with non-organic material as is typical practice in landfills, the amount that can be captured is reduced. For example, the best capture rate proposed in the Bellville project is still only 70 per cent (compared to 83 per cent in Durban), but with separated organic material, this amount gets much closer to 100 per cent. Thus to try to capture methane from a regular landfill, as is the aim of this CDM project, is 'an inefficient solution to an avoidable problem' according to Loots.[19]

It is curious that a project deemed an 'inefficient solution to an avoidable problem' by the very expert in waste management who designed the project, should also be considered to make a 'significant contribution in terms of the local, regional and global environment' under the gold standard. The fact is, as Loots is only too ready to admit, the city of Cape Town simply does not have the resources to institute a large-scale recycling and waste separation scheme. For Loots, 'our first priority is equitable service delivery. We are getting lots of pressure to have a better recycling program and I would love to have a wet/dry program. But it is simply politically unacceptable for recycling to happen in richer neighbourhoods while there is still no roadside collection of waste in poorer ones.'[20]

Thus the argument for awarding South African landfill CDMs with a gold standard label is not that gas extraction is the most sustainable solution, but rather that it is the only approach municipal authorities can afford in the light of present political considerations. Yet this conclusion only reinforces the failure of imagination in the carbon market to produce forward-thinking projects – such as zero-waste strategies – that have long-lasting social and environmental benefits for the community.

A CDM project that provided the capital for a municipality to put in a widespread recycling and waste separation system would have undeniable environmental and social benefits. The space required for landfills would be vastly reduced and without the organic material rotting, they would cause much less nuisance to surrounding areas. In addition to improving productive methane capture from the sorted organic material, the better solution for avoiding climate change, the very act of sorting this would create thousands of employment opportunities, the importance of which cannot be denied in a country like South Africa with an estimated unemployment rate of over 40 per cent.

Surely this is the type of project that a gold standard for the CDM should be certifying? Instead, CDM authorities have chosen to certify a project that provides no employment gains or other social benefits and only further entrenches an unsustainable form of waste management. As such, the gold standard seems vulnerable to the very scourge it was set up to avoid: the propensity of Northern governments to only invest in projects that offer maximum return on investment with little regard for added environmental and social benefits.

Yet as another CDM project in Cape Town displays, even if a gold standard project is able to provide all of the social and environmental benefits it promises, the global carbon market has developed in such a perverse way that it would be unable to make it financially viable.

The Kuyasa low-cost housing energy upgrade project

The CDM executive board officially certified the Kuyasa low-cost housing energy upgrade project as the first gold standard project to receive certified emissions reductions credits. It was a great day for the project developers: the city of Cape Town and SSN, as well as the ten beneficiaries of the project living in Kuyasa, a neighbourhood in the township of Khayelitsha outside of Cape Town. In addition to being a groundbreaking CDM project for Africa and the gold standard, the Kuyasa CDM is the only African project the authors are aware of where the local residents actively supported the project, rather than opposed it (as is the case with the landfill gas capture) or at best were indifferent.

As such, Kuyasa has been held up as an example of the enormous potential of carbon trading to both fight climate change and improve living conditions in local communities.

However, at the time of research (2005–07), the reality of the situation was the opposite; rather than being an example of what the CDM can deliver, Kuyasa is a testament to what it cannot.

On the face if it, there is very little not to like about the Kuyasa CDM project. During the first phase of the project that commenced in July 2002, a handful of small township homes were retrofitted with insulated ceilings, low-watt compact florescent bulbs and solar water heaters on their roofs. In the absence of the water heaters, residents would eventually use electric geysers to heat their water and thus the project creates a hypothetical suppressed demand for coal-fired electricity. The result, in theory, is that 2.85 tonnes of CO_2 per household per year are avoided (SSN n.d. 'Kuyasa Low-Cost Urban Housing'). Ensuring the accuracy of this figure was one of the aims of the first phase of the project where much emphasis is on monitoring the 'baseline methodologies'. The second phase of the project hopes to replicate the baseline study in 2 300 homes.

One of the most likeable aspects of this project is that from the very beginning there have been extensive consultations with the community. The city of Cape Town and SSN have worked closely with the ward development forum in Kuyasa, which put together a broad-based steering committee of community members who were able to take ownership of the project through key decisions. These decisions included assisting with the design of the project, deciding which households would participate and how to move forward into phase two of the project. The steering committee also played an active facilitation role between the project developers and broader community so there were ongoing opportunities for public input over the project. In terms of the gold standard, this project 'attains positive scores in all of the pillars' (SSN n.d. 'Kuyasa Low-Cost Urban Housing').

Financial sustainability?

As wonderful as this project appears to be, when one begins to look into the financial aspects of it, the unfortunate reality of the carbon market is revealed. Of the total budget for the first phase of this project, carbon finance covered only 15 per cent of the upfront costs. SSN officials admit that 'funding is not sustainable'.[21] With carbon credits making up only a fraction of the budget, this project has been able to go ahead because of R12.4 million from the Department of Environmental Affairs and Tourism, another R4 million from the province of the Western Cape and R450 000 from Électricité de France as part of their corporate social responsibility campaign.[22] In addition to this funding, SSN and the city of Cape Town also donated hundreds of hours of labour that were not compensated through project finance.

In September 2005, SSN secured additional funding of R25 million from the provincial administration of the Western Cape and the national Department of Environmental Affairs and Tourism. It is anticipated that this funding will be sufficient to retrofit all 2 299 remaining houses within the project boundary. Much like the first phase, the carbon revenue stream comprises around 15 per cent of the upfront costs of the project at current carbon prices. To this end, the first 10 000 Certified Emissions Reductions (CERs) from this project were sold at a price of €15 to the UK government to offset the G8 summit at Gleneagles (SSN n.d. 'Kuyasa Low-Cost Urban Housing': 6).

With the vast majority of funding for this project now secured through the government, SSN is officially referring to the project as 'a public sector project, relying on government grant funding for its implementation' (SSN n.d. 'Kuyasa Low-Cost Urban Housing': 6). Although the government's support of the Kuyasa project is laudable, it is unlikely to become sustainable financial model, especially in the light of more urgent priorities such as Cape Town's housing backlog of 281 000 units.

The Kuyasa project speaks to the central contradiction in the South African carbon market: projects that are most attractive to investors seem to have the least to offer local communities and vice versa. Even in the less regulated voluntary offset market, this contradiction seems pervasive, as one more case study below will describe.

Careless distribution of light bulbs

As argued more fully in Trusha Reddy's (2006) analysis, light bulbs were not the ideal offset for emissions, given the difficult context in which they were proposed. In lock-step with the increased size and popularity of the CDM has been the voluntary offset market that focuses more on individuals and firms wishing to offset their own emissions in the absence of any legal requirement to do so. This form of trading has reached near celebrity status in recent years with mainstream pop bands such as Coldplay and climate crusaders like Al Gore claiming carbon neutrality by voluntarily offsetting their emissions.

South Africa has not been immune to this trend, although it is unclear whether the voluntary market will prove any more beneficial to the country than the CDM. One story of offsets that illustrates this point involves Climate Care, a British company, who began distributing free energy-efficient light bulbs to low-income South Africans in 2005, via a Cape Town consultancy. Recruited off the street for the distribution job, after two years of being unemployed, was Sibongile Mthembu, 24. Mthembu is a lifelong resident of Guguletu, a sprawling township 20 kilometres from Cape Town, created during the apartheid era.

The idea was that Mthembu would help residents replace the wasteful incandescent variety of light bulbs. After having bought the bulbs (and convinced the city of Cape Town to pay to distribute them) Climate Care was in a position to sell the CO_2 emissions estimated to have been saved to British consumers and companies who wanted to offset their own carbon emissions.

The neighbourhoods where Mthembu went about his ten-day temporary job were full of longstanding problems. Houses were crumbling, with faulty wiring, unpainted ceilings and damp walls. High unemployment and bureaucratic backlog and capacity issues prevented much improvement in most residents' living conditions. 'Some people are pensioners,' explained Pat Mgengi, one resident. 'They don't even get that amount of money every month. They tried taking people out of the houses and we put them back. Even after paying the full amount asked some don't have the title deeds. We are going to court time and again. We are just trying to live like any other human being.'

In this community, the light bulbs Sibongile Mthembu offered would not ordinarily be on anyone's shopping list. At 15 watts, the compact fluorescent bulbs are far more energy efficient than traditional higher-wattage bulbs and last about ten times longer. But they cost US$2.80 each, as opposed to traditional incandescent bulbs at 50 cents, and are not sold locally.

Not surprisingly, Mthembu's bulbs had many takers. Mgengi said he accepted the four that he was offered simply because they were free. 'We just accept what they introduce to us.' But few local people will be able to afford to buy replacements and when asked by residents if he would come back to deliver more bulbs if any were broken, Mthembu admits, he and his fellow light bulb distributors had to lie. Of the 69 low-energy bulbs reported as broken from the households surveyed by Climate Care two months after the project started, none has yet been replaced.

Climate Care argues that this project is generating real carbon savings, since it would not have gone ahead without the firm's intervention and is 'not required by legislation, not common practice (and) not financially viable without carbon funding'. However, in the wake of electricity blackouts, power generator Eskom recently decided to provide five million free energy-efficient light bulbs to low-income households, among a host of other energy-saving measures. Sibongile Mthembu is now employed delivering Eskom's energy-efficient light bulbs to 86 000 houses in Guguletu.

These are houses that Climate Care missed out on its ten-day sojourn in Africa in 2005 and that were supposedly not going to receive such bulbs without Climate Care's money. The problem of course is not Eskom's pledge to provide free light bulbs, but rather the short-sightedness of Climate Care's project implementation and, more importantly, the potentially dangerous legitimisation that projects like this offer Climate Care's customers.

Among Climate Care's biggest customers for its carbon credits are British Airways and British Gas, both major contributors to climate change. British Gas has recently been in the news for pursuing legal action against Bolivia for taking a democratic decision to nationalise its oil resources. It is currently a partner in two large gas fields in the country and has eight exploration blocks that have not yet started production. British Airways, meanwhile, is busy promoting British airport expansion, ramping up its inter-city commuter flight services and launching a budget airline to popular short-haul holiday destinations. Yet Climate Care defends both companies as being among the 'best environmental performers'. 'The climate crisis is so urgent that we should not worry about the motivation of our clients,' the company declares in its 2004 annual report (Climate Care 2004). This position may be acceptable for a voluntary offset company, but one must wonder whether those persons most vulnerable to climate change impacts, such as the few recipients of Mthembu's light bulbs, would agree.

Civic engagement in the carbon market

Outside of the frontline communities living next to or in CDM projects, there are a number of other actors in the South African carbon market, which can be divided into three broad categories. Firstly, consider private sector developers, those 'true believers' in the CDM whose central concern is reducing barriers to easy access of carbon finance. The second group are ENGO reformers, who recognise more serious problems in the carbon market, although they believe that these can be solved through the right mix of policy reform and oversight. Finally, there is an international network that views carbon trading as inherently flawed and believes alternative solutions should be pursued. This group does not believe the problems inherent in the carbon market can be fixed by marginal adjustments, but instead justifies a complete rethinking of our approach to fighting climate change and North-South relations.

Interestingly, among even the most ardent supporters of the CDM there is a sense that all is not well in the South African carbon market. The problem, according to people like project developer Johan Vanderberg of Cape Town-based 'CDM Solutions' is one of institutional capacity. For Vanderberg, it isn't so much the failure of the DNA to provide oversight of the CDM market in South Africa, but rather its inability to process projects quickly enough: 'The biggest issue with the CDM is that it takes a long time; people put a lot of their own money on the line and there are lots of obstacles to overcome. Coming to bank-ability [read: CER purchase agreement] means giving up a pound of flesh in transaction costs.'[23]

Vanderberg estimates that it costs approximately R40 000 to get a project approved and takes a minimum of six months. This cost and time commitment are prohibitive to small-scale producers of either energy efficiency or renewable energy. Even if project developers are able to finance the process and commit the time to getting a project verified, there are still uncertainties about whether the project will be approved and how much they can sell the carbon credits for.

Although the government cannot set the spot market price of carbon to address this latter concern, project developers argue that it could increase the efficiency of the approval process, which will reduce both the time and the costs involved. 'A fast-track procedure is sorely needed,' Vanderberg argues. 'There should be a prima facie view that a CDM project is environmentally beneficial.'[24] The suggestion is that since projects already reduce greenhouse gas emissions, the DNA's sustainable development indicators are unnecessary and inefficient bureaucratic red tape.

The idea of requiring a gold standard or similarly applied benchmark is also considered redundant. For Vanderberg, making sure CDM projects have gold standard validation 'is like saying to a guy with a heart transplant, if this doesn't take away the wrinkles on your face you can't get a new heart'.[25] Palpable results of the project developers' lobbying in this area will be seen in future projects that are approved and policy changes undertaken. To listen to the DNA's Luwazikazi Tyani speak about broadening the sustainability criteria to approve of more projects, it appears the government is only too ready to co-operate with project developers.[26]

For the vast majority of ENGOs in South Africa the problems associated with the carbon market run much deeper than project developers would like to think and thus require much more creative and engaged solutions. While properly representing the views of this broad community is difficult, the positions taken by SACAN seem to be somewhat representative of the various opinions. SACAN has taken a number of strong positions on the CDM and since it is a network of sixteen ENGOs across South Africa, it is fair to assume that many of these positions are widely held in the non-profit sector.

One of the first key differences between these ENGOs and the project developers is their comprehension and in some cases sympathy with the ideological critiques against carbon trading. In the July 2002 edition of *Climate Action News*, SACAN's quarterly newsletter disseminated throughout South Africa – the headline of the front cover story on the CDM read 'Can we justify selling Africa's atmosphere?' The fire line of this story was even more to the point on the ideological critique of carbon trading (SACAN 2002: 1).

While appreciating some of the theoretical critiques, ENGOs see the injustices of the CDM most clearly in some of its more controversial projects. As has previously been discussed, SACAN played a key role in highlighting the lack of additionality in the Sasol project and may well be responsible for the company's decision to withdraw the project in its original form.

The other project that is widely opposed by ENGOs in South Africa is Bisasar Road. For SSN, which is an active member of SACAN, the cause of all the problems in Durban lies with the involvement of the PCF and the World Bank.[27] While SACAN has found plenty of problems with the CDM and seems to sympathise with the claim that free market economics contributes to this, the organisation doesn't believe that the two forces must be addressed simultaneously. SACAN leader Richard Worthington thinks that rejecting Kyoto on the basis of its market logic is misguided, 'Sure, I'd love it if we had a more co-operative economic system in place, but we can't wait for that before tackling climate change . . . [This is] a poor strategy that plays into the hands of Bush.'[28] Thus a more reformist approach to the problem is adopted, where SACAN attempts to influence CDM development by shaming bad projects, supporting better ones, and advocating for strong reforms so there is more to support and less to shame.

While SACAN has shied away from attacking the market logic of carbon trading, a growing network of climate justice activists have no such reservations and indeed see it as their contribution to the climate change issue. Known as the Durban Group for Climate Justice, after the city that hosted their founding conference and the name of their founding declaration, these activists reject the claim that 'carbon trading will halt the climate crisis'. Rather, this group argues that the crisis is caused by the mining and use of fossil fuels, something that carbon trading fails to address and in many ways solidifies, thus making it a 'false solution which entrenches and magnifies social inequalities'.[29]

South African participants at the founding conference in October 2004 have almost exclusively confined their strategic actions to the struggle around Bisasar Road and have shown little awareness and less engagement with other CDM projects. The activities around Bisasar Road include op-ed pieces in national media and the production of a short film on the subject for the South African Broadcasting Corporation (SABC) by local filmmaker Rehana Dada. Amsterdam-based Carbon Trade Watch and other organisations involved with the Durban Group for Climate Justice also issued a public letter to the PCF articulating their concerns over the lack of consultations on the CDM project and its entrenching of environmental injustices in the community. Finally, until her death in July 2007, Khan continued her courageous court battle against the landfill, although it should be noted that this commenced long before the Durban Group for Climate Justice was formed.

Although it seems premature to judge the impact of the Durban Group for Climate Justice, the non-governmental organisations (NGOs) and academics do seem to be more effective in the international policy arena than at the local level. With most mainstream environmental NGOs no longer engaging in debates on carbon trading, the Durban Group for Climate Justice can be lauded for being the only voice for people affected by CDM projects at the annual Conference of Parties (COP) and other international gatherings. Yet for the time being, this has yet to translate into successful local campaigns against carbon trading projects in Southern countries. South Africa is no exception to this and apart from the Bisasar Road project, signatories to the Durban declaration in South Africa have had less impact on their own carbon market than the more moderate ENGOs who are not signatories.

Conclusion

It appears that many of the troubling trends apparent in the global carbon market are being replicated in South Africa. These include dubious projects adversely impacting on local communities, profit-oriented private sector developers neglecting additionality and renewable energy projects – that were pitched as the centrepiece of the CDM – remaining unsustainable on carbon finance alone. At an institutional level, compromised and/or under-resourced civil servants are unable to address these contradictions. Some social actors can articulate this critique, yet the actual struggle against projects on the ground continues to be in isolation, with few activists taking the time to assist communities wanting to engage in this issue and access its technical jargon and complex processes.

Alternatives to this sobering status quo are not hard to find. The issue is less one of accepting the merits of community-based renewable energy and more about recognising the power imbalances in geopolitical relations. Neither carbon trading nor its incarnation in the South African carbon market exists because they are a preferable solution to the climate crisis, at least from the perspective of poor South Africans on the front lines of climate vulnerability. Rather, this system is a product of capital interests and none loom larger in the climate issue than the all-powerful fossil-fuel industry. While combating this enemy is a much more daunting task than tweaking additionality criteria, it may prove the only acceptable solution to the greatest number of people at risk.

Notes

1. For up-to-date information on the status of the South African carbon market, please consult the Designated National Authority (DNA) website: http://www.dme.gov.za/dna/index.stm.
2. Interview with Lindsay Strachan, engineer at Durban Solid Waste, 13 June 2005.
3. For more on the history of Clare Estates and the Bisasar Road landfill, consult Reddy 2005a.
4. Interview with Lindsay Strachan, engineer at Durban Solid Waste, 13 June 2005.
5. Ibid.
6. Ibid.
7. Ibid.
8. Ibid.
9. Interview with senior engineer at Durban Solid Waste, 13 June 2005.
10. Interview with Lindsay Strachan, engineer at Durban Solid Waste, 13 July 2005.
11. Ibid.
12. Private correspondence with Gerrit Kornelius, 22 July 2005.
13. This figure is somewhat controversial, at least to Lindsay Strachan, engineer at Durban Solid Waste, who believes 'they're over-estimating their LFG potential' (interview, 13 July 2005).
14. According to Lester Malgas of SSN, 'Durban's perfume rods (used to offset the rotting stench of garbage) leave a bad taste in everyone's mouth' (interview, 30 June 2005).

15. Interview with Deborah Roberts, 28 July 2005.
16. Interview with Sheriene Rosenberg (SSN), 30 June 2005.
17. The issue of the gold standard is discussed in further detail below. For more information, see: www.cdmgold standard.org.
18. Interview with Walter Loots, head of Solid Waste (Cape Town).
19. Ibid.
20. Ibid.
21. Interview with Lester Malgas.
22. Interview with Sheriene Rosenberg (SSN), 30 June 2005.
23. Interview with Johan Vanderberg, 13 July 2005.
24. Ibid.
25. Ibid.
26. Interview with Luwazikazi Tyani, 28 June 2005.
27. Interview with Sheriene Rosenberg (SSN), 30 June 2005.
28. Interview with Richard Worthington (SACAN).
29. Durban Declaration on Climate Justice, Appendix 2 in this volume and online: www.carbontradewatch. org/durban.

PART 3

WHO REALLY BENEFITS FROM CARBON TRADING?

4

Climate Fraud and Carbon Colonialism

Heidi Bachram[1]

'The rush to make profits out of carbon-fixing engenders another kind of colonialism.' (Centre for Science and the Environment 2000)

In order to understand the impact of pollution permits and emissions trading on the ecological crisis, the findings of the international scientific community must be noted.[2] The Intergovernmental Panel on Climate Change (IPCC), a United Nations (UN) advisory body numbering 3 000 scientists, concluded in 2001 that 'the present CO_2 [carbon dioxide] concentration has not been exceeded during the past 420 000 years and likely not during the past 20 million years'.[3] The clear and alarming consensus in the scientific community is that humankind is wreaking havoc on the atmosphere. Across the world 80 million people are at severe risk of their homes and livelihoods being destroyed by flash flooding as sea levels rise, fed by melting icecaps and extreme weather events are becoming more frequent. Although these weather changes will occur everywhere, poorer countries will have less ability to adapt. Meanwhile the emissions of greenhouse gases, which are creating the problems, come overwhelmingly from the richer industrialised countries that do have the resources to adapt. For example, the United States and the European Union, with only 10 per cent of the world's population, are responsible for producing 45 per cent of all emissions of CO_2, the principle greenhouse gas.[4]

Three-quarters of all the CO_2 emitted by human activities is from burning fossil fuels (IPCC 1995). The rest mostly comes from deforestation. The IPCC concludes that an immediate reduction of 50–70 per cent of CO_2 emissions is necessary to stabilise the concentrations in the atmosphere. In its most recent report, the IPCC states that 'eventually CO_2 emissions would need to decline to a very small fraction of current emissions'. Faced with this looming climate crisis, the global community of states' response has been the passing of the Kyoto Protocol in 1997, slowly ratified by 156 countries and infamously rejected by the world's biggest polluter – the United States. At the core of the Protocol is an agreement to reduce emissions by an average of 5.2 per cent below 1990 levels of greenhouse gases by the year 2012. Larry Lohmann vividly sums up the inadequacy of this agreement: 'Shortly after the treaty was initialled in 1997, a scientific journal pointed out that 30 Kyotos would be needed just to stabilise atmospheric concentrations at twice the level they stood at, at the time of the Industrial Revolution. At this rate, 300 years of negotiations would be required just to secure the commitments necessary by the end of this decade.' (2002)

Also agreed upon in 1997 was the main mechanism for achieving this target, tabled by the United States in response to heavy corporate lobbying: emissions trading. This market-driven mechanism subjects the planet's atmosphere to 'legal' emissions of greenhouse gases. The arrangement parcels up the atmosphere and establishes the routine buying and selling of permits to pollute, as though they were like any other international commodity. The National Institute of Public Health in the Netherlands (RIVM) estimates that with emissions trading, the actual reductions achieved in terms of the Kyoto Protocol will only be 0.1 per cent, far below the already inadequate 5.2 per cent reduction from 1990 levels (RIVM 2001).

In addition, as we shall show, emissions trading is rife with controversy and the potential for exacerbating environmental and social injustices. The changes necessary to avert climate catastrophe are simple enough, namely, a switch away from fossil fuels and to renewable energy sources such as solar and wind, along with a reduction in energy use generally. Instead, world leaders have taken ten years to agree to inadequate targets and the deeply flawed mechanism of emissions trading. Although emissions trading is represented as part of the solution, it is actually a part of the problem itself. Despite the scope and gravity of the dangers posed by greenhouse gases and the major role of emissions trading in compounding them, this arrangement has not been seriously challenged in any international forum. The continuing acquiescence toward emissions trading is not an accident or bureaucratic oversight. The smooth sailing of this arrangement is attributable to the arm-twisting tactics of the richer nations and their constituencies of corporate polluters whenever global treaties are hammered out. The failure of the Kyoto Protocol to deal adequately and effectively with climate change is also representative of wider issues of democratic decision-making and symptomatic of the injustices that permeate international relationships.

What is emissions trading?

Under the Kyoto Protocol, the polluters are countries that have agreed to targets for reducing their emissions of gases in a defined time period. The polluters are then given a number of emissions credits, equivalent to their 1990 levels of emissions minus their reduction commitment. These credits are measured in units of greenhouse gases, so 1 ton of CO_2

would equal one credit. The credits are licences to pollute up to the limits set by the commitment to achieve the average reduction of 5.2 per cent agreed in Kyoto. The countries then allocate their quota of credits on a nation-wide basis, most commonly by 'grandfathering', so that the most polluting industries will receive the biggest allocation of credits (IETA 2001). In this system it pays to pollute. Several possibilities then exist:

- The polluter does not use its whole allowance and can either save the remaining credits for the next time period (bank them), or sell the credits to another polluter on the open market;
- The polluter uses up its whole allowance in the allotted time period, but still pollutes more. In order to remain in compliance, spare credits must be bought from another polluter that has not used up its full allowance;
- The polluter can invest in pollution reduction schemes in other countries or regions and in this way earn credits that can then be sold, or banked, or used to make up shortfalls in its original allowance.

Credit-earning projects that take place in a country with no reduction target (mostly in the developing world) come under the contentious rubric of the Clean Development Mechanism (CDM). There have already been signs that traditional Overseas Development Aid (ODA) given by developed countries will be used to fund CDM projects. Instead of building wells, rich countries can now plant trees to offset their own pollution. Projects that take place in countries with reduction targets come under joint implementation (JI). For example, an energy efficiency program in Poland funded by a UK company could qualify. It appears that JI projects will mainly take place in Eastern Europe and Russia, where equivalent reductions can be made more cheaply as costs and regulatory standards are lower.

Both CDM and JI projects can be of different kinds: monoculture tree plantations, which theoretically absorb carbon from the atmosphere (carbon sinks); renewable energy projects such as solar or wind projects; improvements to existing energy generation; etc. The amount of credits earned by each project is calculated as the difference between the level of emissions with the project and the level of emissions that would occur in an imagined alternative future without the project. With such an imagined alternative future in mind, a corporate polluter can conjure up huge estimates of the emissions that would be supposedly produced without the company's CDM or JI project. This strategy allows for a high (almost limitless) number of pollution credits that can be earned for each project. It allows the company to pollute more at other sites, to sell its credits to other polluters, or to engage in a combination of these lucrative tactics. Its long-term consequences are increased greenhouse gas emissions and increased corporate profit obtained from their production.

There is yet another provision in emissions trading that introduces increasing levels of complexity and confusion: the pollutants are interchangeable. In effect, a reduction in the emission of one greenhouse gas (e.g. CO_2) enables a polluter to claim reductions in another gas (e.g. methane). Thus, progress in cleaning up the atmosphere might appear to be going forward, while closer scrutiny reveals that no actual improvement is taking place.

Climate fraud

While many hundreds of millions of dollars are being invested in setting up emissions trading schemes all over the world (the UK government alone has spent £215 million on its trial trading scheme) virtually no resources are being channelled into their regulation. This imbalance can only lead to an emissions market dangerously reliant upon the integrity of corporations to file accurate reports of emissions levels and reductions. In practice, corporations such as PricewaterhouseCoopers are acting as both accountants for and consultants to polluting firms, as well as verifiers of emission reduction projects. Some entrepreneurial firms such as CH2M Hill and ICF Consulting are also offering consultancy and brokerage, as well as verification services. These potential conflicts of interest were at the heart of scandals relating to Enron and Arthur Andersen who were both pioneers in emissions trading.

Opportunities for fraud abound as the poorly regulated emissions markets develop. This is inevitable in the laissez-faire environment in which emissions trading is conducted. In the first year of the United Kingdom's trial emissions trading scheme in 2002, Environmental Data Services (ENDS) exposed the main corporations involved in the scheme as having defrauded the system. They found that three chemical corporations had been given over £93 million in 'incentives' by the UK government for their combined commitments to reduce pollution by participating in the voluntary trading scheme. However, the corporations had already achieved their promised reductions under separate compulsory EU-wide regulations. ENDS estimated that one corporation, DuPont, could make a further £7 million from the market value of the 'carbon' credits generated (ENDS 2003). Therefore the corporations had received millions of UK taxpayers' money for doing nothing. This was only highlighted by the independent work of the ENDS service and no government monitoring of the scheme revealed these instances of fraud. No subsequent action was taken by the government to respond to these revelations.

Monitoring the monitors

At present, there is no consensus on the international monitoring of emissions trading or the means to verify claimed reductions in greenhouse gas emissions. The prospects for such monitoring and verification are still under discussion. Nevertheless, hundreds of credit-generating projects are going ahead and at least three EU countries (Denmark, the Netherlands and the United Kingdom) have begun their own internal greenhouse gas trading schemes, with an EU-wide market beginning in 2005. What has been emerging in place of UN or government-led guidance are initiatives taken by non-governmental organisations (NGOs), corporate-led self-monitoring and entrepreneurial verification schemes by consulting firms.

Environmental non-governmental organisations (ENGOs) such as the Worldwide Fund for Nature (WWF) are developing labelling standards for CDM projects, similar to other controversial labelling schemes such as the Forest Stewardship Council (FSC) accreditation.[5] Alongside this, more critical NGOs such as SinksWatch, the World Rainforest Movement (WRM) and CDMWatch attempt to monitor trades and support communities affected by projects by providing them with crucial research and campaigning tools. However, these latter groups are often poorly funded and under-resourced and it is impossible for NGOs to systematically monitor the thousands of transactions that are expected to take place globally once the greenhouse gas markets come into being.

Meanwhile, oil giants BP and Shell have been experimenting with internal trading schemes and have employed self-monitoring to report trades and verify reductions. There are obvious conflicts of interest affecting the reliability of data produced in this way. For example, BP states that its internal trading scheme achieved a 5 per cent reduction in CO_2 emissions, half of its voluntary commitment of 10 per cent reductions below 1990 levels. The scheme also earned the company US$650 million in extra profits, as most reductions were achieved through energy efficiency and reducing gas flaring. BP admitted that measuring reported emissions is 'never 100 per cent accurate' (Akhurst 2002). However, there is no independent corroboration for these figures as the data was monitored internally by BP itself.

Lastly, consulting firms such as Det Norske Veritas (DNV) have taken up the verification of emissions reductions. In 2002, for instance, DNV validated a eucalyptus plantation, a project funded by the World Bank's Prototype Carbon Fund (PCF). The plantation is the target of local and international campaigns, as monoculture eucalyptus causes severe problems for local people and the environment. While admitting in its report that it could not guarantee that the carbon would be permanently stored in the plantation, DNV nonetheless recommended the project to the CDM executive board (DNV 2002).[6]

There are serious concerns about the effectiveness and wisdom of relying upon any of these monitoring and verification practices, yet a reliable surveillance system is essential to prevent the Kyoto targets from being undermined by fraudulent and destructive projects. However, it is difficult to imagine how any organisation, UN-sanctioned or otherwise, could cope with the vast amount of trade that will take place globally.

Carbon colonialism

The Centre for Science and the Environment in India observes that so-called carbon-fixing projects are in reality opening the door to a new form of colonialism, which utilises climate policies to bring about a variation on the traditional means by which the global South is dominated (see Centre for Science and Environment 2000). In particular this trend is seen in the use of monoculture plantations, which allegedly sequester or remove CO_2 from the atmosphere. Scientific understanding of the complex interactions between the biosphere (trees, oceans, etc.) and the troposphere (the lowermost part of the atmosphere) is limited. Furthermore, there is scientific consensus that the carbon stored above-ground (i.e. in trees) is not equivalent to the carbon stored below-ground (i.e. in fossil fuels). Therefore there is no scientific credibility for the practice of soaking up pollution using tree plantations (for a more detailed discussion of this, see Lohmann 2001; 2005). Yet entrepreneurial companies such as Face International are charging ahead with plantations, while propagating the idea that consumers need not change their lifestyles. This new logic dictates that all that needs be done to become carbon neutral is to plant trees. The majority of these projects are being imposed by the North upon the South.

The key questions revolve around whether the concept of carbon offsetting is either tenable or desirable. The various schemes of CDMs and JI mechanisms of the Kyoto Protocol rely on the notion that emissions from a polluting source can be nullified through investments in renewables or carbon sinks. These compensation mechanisms vary in complexity and design, but all are enthusiastically promoted by the emerging offset industry, which is being developed to service the new markets. As a result, clients wishing to be carbon neutral are bombarded with a plethora of new, untested and poorly thought-through offset products and services.

Companies such as Future Forests sell branded carbon offset products to promote so-called CarbonNeutral™ living. They offer a consumer the possibility to take CarbonNeutral™ flights, go CarbonNeutral™ driving, live in CarbonNeutral™ homes and be a CarbonNeutral™ citizen, by planting trees that theoretically absorb carbon from the atmosphere.[7] The gathering of global business elites, the World Economic Forum, promotes its events as CarbonNeutral™ with the aid of these self-styled offset businesses. The allure of offset culture is understandable. Corporations, ever conscious of cost and image, seek quick-fix solutions that do not require radical changes to fundamental business practice.

However, there are many problems with this approach. Offset schemes typically do not challenge the destructive consumption ethic, which drives the fossil-fuel economy. These initiatives provide moral cover for consumers of fossil fuels. The fundamental changes that are urgently necessary, if we are to achieve a more sustainable future, can then be ideologically redefined or dismissed altogether as pipe-dreams. Furthermore, land is commandeered in the South for large-scale monoculture plantations, which act as an occupying force in impoverished rural communities dependent on these lands for survival. The Kyoto Protocol allows industrialised countries access to a parcel of land roughly the size of one small Southern nation – or upwards of 10 million hectares – every year for the generation of CDM carbon sink credits.[8] Responsibility for over-consumptive lifestyles of those in richer nations is pushed onto the poor, as the South becomes a carbon dump for the industrialised world.[9]

On a local level, longstanding exploitative relationships and processes are being reinvigorated by emissions trading. Indigenous communities, fisherfolk and other marginalised rural Brazilian peoples were systematically removed from land during the colonial obsession with plantations. Now the World Bank is funding a eucalyptus plantation in Brazil run by an existing plantation company called Plantar, with the intention that it be approved as a CDM project. While plantations have their own ecologically destructive qualities such as biodiversity loss, water table disruption and pollution from herbicides and pesticides, their social impact is equally devastating for a local community. Lands previously used by local peoples are enclosed and in some cases they have been forcibly evicted. This was the case in Uganda when a Norwegian company leased lands for a carbon sink project, which resulted in the eviction of 8 000 people in thirteen villages (WRM 2000).

The workers on such plantations have little or no health and safety protection and are exposed to hazardous chemicals and dust particles. Plantar is a company with an especially sordid history. In March 2002 the regional labour office prosecuted 50 companies, among them Plantar, for the illegal outsourcing of labour, a process synonymous with extreme exploitation. Indeed, in the 1990s, the Montes Claros Pastoral Land Commission, an organisation originating in the Catholic Church and well-respected in the region, verified that slave labour was used on the company's property (WRM 2002).

Similar disregard exists for the natural environment. Thus local fisherfolk in the regions around the plantations in Brazil are poverty stricken due to the pollution caused by the overuse of pesticides and herbicides, which contaminates rivers and water sources and kills fish. In some cases, the water in streams and rivers has entirely dried up because the non-indigenous eucalyptus is a thirsty tree. With the World Bank's assistance, this plantation will now expand by 23 400 hectares. This is a disaster for local agriculture and people dependent on water sources for subsistence. The ruination caused by the trafficking in pollution credits serves only to place the cloak of ecological respectability over local and global unequal power relations.

Might makes right

One of the more tragic ironies of the Kyoto Protocol is that carbon sinks (forests, oceans, etc.) can only qualify for emission credits if they are managed by those with official status.

This means that an old-growth rainforest inhabited for thousands of years by indigenous people does not qualify under Kyoto rules and cannot get credits. However, a monoculture plantation run by the state or a registered private company does qualify. This exposes the vested interests that are served by emissions trading, as ordinary people are not recognised by the official process. Neither does the Kyoto Protocol offer protection for forests. Instead emissions trading provides an opportunity for extended encroachment on the lives of indigenous peoples by governments and corporations, expanding the potential for neocolonial land-grabbing. Furthermore, other ecosystems such as grasslands are not protected under the Kyoto Protocol and therefore a monoculture plantation could supplant them. Under the guise of creating solutions for one environmental problem, climate change, further destruction of diverse ecosystems has been legitimised.

Emissions trading represents the latest strategy in an ongoing process that stems from sixteenth-century European land enclosures to the recent World Trade Organisation (WTO) negotiations on public health and education to privatise and liberalise the global commons and resources. By its very nature, an emissions credit entitles its owner to dump a certain amount of greenhouse gases into the atmosphere. Control of such credits effectively leads to control of how the atmosphere, perhaps the last global commons, is used. The Kyoto Protocol negotiations have not only created a property rights regime for the atmosphere, they have also awarded a controlling stake to the world's worst polluters, by allocating credits based on historical emissions. A similar relationship applies to the process leading to the agreements in Kyoto.

The 1992 Rio de Janeiro Earth Summit

From the beginning of international discussions about climate change, Northern governments and corporate polluters have been opposed to the structural changes needed to truly combat the problem. Before the Rio Earth Summit, an International Negotiating Committee (INC) was set up to formulate a draft text. Within the INC, both the United States and the European Union argued against binding reductions in greenhouse gas emissions (Halpern 1992). The Earth Summit did however produce the United Nations Framework Convention on Climate Change (UNFCCC). Despite some obvious merits, such as the recognition that climate change was an urgent issue for the first time in an international agreement, the UNFCCC did not include any commitment to legally binding emission reductions. Nor did it recognise the role of industry, overconsumption and free trade policies in exacerbating climate change.

Meanwhile in 1991 the United Nations Conference on Trade and Development (UNCTAD) set up a department on the trade in greenhouse gases. Emissions trading then found its way onto the INC's agenda at its third session held in Nairobi in September 1991.

UNCTAD also set up the International Emissions Trading Association (IETA), a corporate lobby group dedicated to promoting emissions trading. These activities led to a May 1992 report entitled 'Combating Global Warming: Study on a Global System of Tradable Carbon Emission Entitlements', produced with financial support from the governments of the Netherlands and Norway. The intimate connections between business and the UN is further evidenced in that the former head of UNCTAD's emissions trading division, Frank Joshua, is now the global director for greenhouse gas emissions trading at Arthur Andersen.

Formal proposals for trading emissions, however, were not made until the mid-1990s. By then UNCTAD's research on greenhouse gas trading was well advanced; it never pursued research into other alternatives, or even other market-based instruments, such as pollution taxes. The neoliberal bias of the UN in this instance seems less a question of succumbing to corporate pressure than of an organisational culture oriented towards corporate-friendly solutions as a matter of course.

The role of corporations

Corporate lobby activity before the Rio Earth Summit remains to be researched, but it is telling that most of industry's goals for the Earth Summit (i.e. promoting 'cost-effective policies' and 'self-regulation') were achieved. Considering the corporate connections to government delegations, it is unsurprising that they were so successful. For example, the chair of the working party on sustainable development in one of the most powerful corporate lobby groups in the world, the International Chamber of Commerce (ICC), was also a member of the UK official delegation in Rio (Beder 1997). The ICC continues to have privileged access to policy-makers and regularly makes statements to the INC on climate change, representing the voice of business. The voices of neoliberal ideology seem consistently to be heard loud and clear in all international forums on climate change.

Corporations also promote business-friendly solutions through partnerships with NGOs, governments and the UN. This tactic is new and exposes some dissension within corporate ranks. Enron, for example, saw that Kyoto 'would do more to promote Enron's business than will almost any other regulatory initiative' and the company was one of the main proponents of emissions trading (Horner 2002). Along with expensive public relations campaigns, such as BP's environmental 'Beyond Petroleum' makeover, so-called progressive corporations have successfully advanced the concept of public-private partnerships (PPPs), wooing NGOs and public opinion with slick public relations campaigns and advertising. This approach was epitomised by what happened at the World Summit on Sustainable Development in Johannesburg in 2002. No legally binding agreements were reached at this second Earth Summit. Instead, over 280 PPPs were showcased, highlighting the lack of political will on the part of governments and the extravagant enthusiasm of corporations for taking control of the issue.

Co-opting NGOs

ENGOs have also been hypnotised by corporate multi-stakeholder dialogues. Part of the formula for developing an image of the good corporate citizen is to enlist the help of friendly NGOs in controversial activities, effectively outsourcing legitimacy. ENGOs can therefore provide a moral stamp of approval for corporations involved in emissions trading. The conflict of interest involved in verifying the emissions of companies who are paying you to do so, while also providing general funding for your organisation, is obvious. 'Working with business is as important to us as munching bamboo is for a panda,' says a WWF representative – unsurprisingly, since WWF receives approximately £1 million a year from corporations in the United Kingdom alone and has an operational budget larger than the WTO (Rowell 2001). Recently WWF stated that emissions trading in the European Union could be an 'important element' in climate policy and help to 'prevent dangerous climate change . . . as cost-effectively as possible' (WWF 2002).

However, it is not only conservative ENGOs that have been neutralised by the strategies of corporate polluters. At the original Earth Summit in Rio the NGO Global Forum produced an alternative treaty, designed to influence the official Rio declarations. In this visionary document, the NGOs declared that the climate negotiators should 'avoid any emission trading schemes which only superficially address climate change problems, perpetuate or worsen inequities hidden behind the problem, or have a negative ecological impact' (NGO Alternative Treaties 1992). After Kyoto, however, the large NGOs that had helped to produce the alternative treaty in Rio de Janeiro began to abandon their stand against emissions trading. By November 2000 at the sixth Conference of Parties (COP 6) of the signatories to the UNFCCC, even some of the more radical NGOs like Friends of the Earth had changed their position on emissions trading. At COP 6 they moderated their demands, calling for a 20 per cent limit on the use of emissions trading. Eight months later, after agreement was reached on key controversial issues in the Kyoto Protocol at the COP in Bonn in July 2001, press statements from Friends of the Earth International heralded the agreement as a 'new hope for the future' – even though it placed no specific limits on the use of emissions trading and was actually weaker than the deal they had described as 'junk' at COP 6.

In Johannesburg at the 2002 World Summit on Sustainable Development, Greenpeace and the World Business Council for Sustainable Development (WBCSD), which includes corporations such as Dow Chemical and General Motors, made a joint declaration on climate change, urging governments to move forward. This happened despite the fact that the

WBCSD still does not necessarily endorse implementation of the 1997 Kyoto Protocol, in sharp contrast to the stated aims of Greenpeace. At the Earth Summit in 1992, Greenpeace and the WBSCD fought like cats and dogs. Ten years later they stood on the same platform, but without a substantial common vision of how governments should move forward.

A number of mainstream NGOs that have long campaigned for an international agreement on climate change are now persuaded that business support is crucial. Part of the reason is technocratic. In the lengthy negotiation process, the talks tend to become extremely technical and the language impenetrable to the point that most people participating do not understand fully the implications of the compromises made. In effect, environmental policy decisions are often left in the hands of so-called 'climate experts' in organisations, with the knock-on effect that democracy and understanding within NGOs suffers and public statements are reduced to simplified slogans. At times, even well-intentioned activists in NGOs are persuaded by the win-win scenario rhetoric that accompanies emissions trading. Talk of 'technology transfer' and 'leapfrogging industrialisation' is seductive. Yet at the heart of this corporate paternalism lies the stone-cold logic of the free market. This has created a situation where the NGO world has been thrown into confusion and discord. While mostly Northern mainstream NGOs support, or do not resist, emissions trading, many social movements and smaller NGOs are vehemently opposed to it. Now that NGOs have been effectively diverted, corporate interests have been placed at the heart of political negotiations and industry has been defined as a legitimate stakeholder.

The impact of the WTO on emissions trading

Proponents of emissions trading argue that as schemes are implemented, the rules governing them can be tightened and improved and fraud avoided. This view is at best naive and at worst dishonest. As emissions trading emerges as the principal component of government climate change policy, the rules for its use will have to conform to the general rules governing trade. Any efforts to improve the rules of emissions trading, or to curb its use, will be subject to the general forces of liberalisation. Industry lobby groups and neoliberal think tanks want WTO compliance across the board, with no exceptions made for other purposes or values. Many corporate lobby groups, in particular, want unrestricted free trade in greenhouse gas credits, rather than government regulation and taxation to achieve emissions reductions (CEO 2000). Since the rules for the Kyoto mechanisms are still being developed and the WTO's committee on trade and environment (the principal committee responsible for evaluating the relationship between multilateral environmental agreements such as the Kyoto Protocol and the WTO) is still deliberating, much remains speculative. However, there are already many areas of likely conflict. The net effect may be to water down regulation of emissions trading in order to avoid trade conflicts.

Environmental justice

A further fundamental problem of emissions trading is its tendency to perpetuate and aggravate environmental injustice. The six greenhouse gases due to be traded all have toxic co-pollutant side effects.[10] This aggravates other dimensions of social injustice in as much as polluting industries are disproportionately located in low-income areas and in communities of colour. A clear example of this is in the case of a sulphur dioxide trading scheme in Los Angeles, RECLAIM (Regional Clean Air Incentives Market), where localised pollution of the local Latino communities around factories involved in the scheme continued unabated (Drury et al. 1999). It is likely that this phenomenon will be widely replicated with global greenhouse gas trading. Reductions will not need to take place at their source, allowing factories to continue polluting locally. The communities affected are those with the least power to resist; pollution ghettoes are thereby created, bringing the seemingly abstract nature of the market into deadly focus (Sandborn, Andrews and Wylynko 1992).

The introduction of emissions trading means that precious time and resources are being channelled away from the solutions that could successfully resolve climate change in a just way. It took ten years to put the RECLAIM programme in place in Los Angeles and the Kyoto market will not officially begin trading until 2008. By then national governments will have spent millions setting up their internal schemes in preparation for the international market. Brokers, consultants, NGOs, corporations, public relations firms, speculators, as well as opportunistic experts and consulting firms that offer science for sale, will be created in anticipation of the new carbon economy. All this energy, investment and time could be put into more positive and effective strategies to resolve climate change and at the same time, to combat environmental injustice. Besides central government measures, from taxation and subsidies to laws, grass-roots initiatives of all kinds could provide answers at low cost, while also successfully tackling issues of environmental injustice and carbon colonialism.

The alternatives

One alternative to corporate-led schemes such as emissions trading is government regulation. This could include taxation, penalties for polluting and imposed technological fixes, such as scrubbers and filters on smokestacks. Such an approach has been successfully adopted in Iceland (where 99 per cent of electricity comes from geothermal sources) and Costa Rica (where 92 per cent of energy comes from renewables). Additionally, government fossil-fuel subsidies and tax breaks could be withdrawn and subsidies for small-scale renewables increased instead. However, there are problems with this approach as well. In Iceland, one of the main producers and distributors of renewable energy is the oil giant Shell. Although the product has changed from fossil fuels to renewables, the corporation is still the same.

The power dynamic remains; often the renewable investments of large fossil-fuel corporations are another tactic in a cleverly planned 'greenwash' campaign to improve their public image.

Furthermore, the failure to challenge corporate monopolies in the renewable energy sector could stifle diversity and innovation, as can be seen when comparing developments in the Netherlands and Germany. In the Netherlands, subsidies for the solar industry in the 1990s were concentrated on Shell and eco-consultants Ecofys. This limited the number of solar panel firms to just a few main players and Shell gained a virtual monopoly in solar panel installation. In contrast, German subsidies were distributed more fairly across different sized firms. By 2002 there were over 300 companies involved in supplying solar panels.[11] Even a future where wind and solar are the main sources of energy still fails to challenge underlying patterns of consumption and does not guarantee that transnational corporations will suddenly behave in an environmentally or socially just way.

Many grass-roots initiatives have nevertheless arisen to tackle these problems and it is here that we can see the outlines of a holistic approach to the problem posed by climate change. Thousands of small-scale projects that successfully balance social and economic injustice with environmental sustainability have already sprung up around the world. The Centre for Alternative Technology in Wales, for example, is in the process of building a wind turbine, a project that was initiated and is managed by the local community. The energy will be used locally and any surplus sold and the dividends shared among the community group.[12] Another initiative is in the process of being launched in northern Spain by a project called Escanda, which is engaged in planning and forming a renewable energy co-operative to invest, build and maintain wind and solar energy. This challenges corporate control of energy production and distribution, promoting empowerment and democracy, as decision-making is held by the people producing and using the electricity generated. It is hoped that the project can provide a model for other communities in Spain and perhaps be applied across Europe.[13]

Another method is employed by Khanya College in Johannesburg, where a community education programme to tackle issues of climate change from an environmental justice perspective is being planned. Community educators and activists will conduct workshops to inform and train township residents in the province on the impacts and effects of climate change upon their lives. The workshops open up a safe political space where the communities can explore the issues and create their own solutions. This unique synthesis of education and empowerment is absent from the official process and diametrically opposed to the top-down solutions offered by proponents of emissions trading schemes. What all these community-based projects have in common is an innovative, yet practical, combination of economics, ecology, democracy and participation.

Conclusion

In the best-case scenario that emissions trading is strictly regulated, it is still unlikely to achieve even the woefully inadequate reductions in greenhouse gas emissions enshrined in the Kyoto Protocol. This would be true even if the United States joined the rest of the major polluting countries in ratifying the Protocol. Yet should a foolproof monitoring system be put in place, the whole system would lose its appeal of being cheap and unchallenging for corporations and so any attempt to introduce such methods will be strongly opposed. Furthermore, the neoliberal trends in international trade make it unlikely that emissions markets will ever be tightly regulated. The strategy and tactics of emissions trading have been adorned with the rationale of neoliberal ideology; they have become so institutionalised in international forums that regulatory initiatives are unlikely to be proposed from within their circles.

Yet even if emissions trading were adequately regulated, the reality is that the trading in pollution best serves the needs of those with the most to lose from resolving the climate crisis. As climate change exposes fundamental flaws in the current world order, only the most challenging responses will have any prospect of success. Transnational fossil-fuel corporations and the governments of industrialised countries will not concede power willingly and hence emissions trading is being used to distract attention away from the changes that are urgently needed. In this way corporations and governments are able to build the illusion of taking action on climate change, while reinforcing current unequal power structures. Emissions trading therefore becomes an instrument by means of which the current world order, built and founded on a history of colonialism, wields a new kind of carbon colonialism.

As with the colonialism of old, this new colonising force justifies its interference through moral rhetoric. As the colonisers seek to resolve climate change, they conveniently forget the true source of the problem. With the looming climate crisis and the desperate need for action, the resulting course recommended by corporations and government is not analysed critically. The debate is transformed, shifting the blame onto the poor masses of the global South. Lost in this discourse is the reality that the world's richest minorities are the culprits who have over-consumed the planet to the brink of ecological disaster. Instead of reducing in the rich countries, a carbon dump is created in the poor countries and thus rich countries can continue in their unequal overconsumption of the world's resources.

On almost every level of emissions trading, colonial and imperialistic dimensions exist. There may be new labels for these phenomena, such as environmental injustice, but the fundamental issues are the same. The dynamics of emissions trading, whereby powerful actors benefit at the expense of disempowered communities in both North and South, is a modern incarnation of a dark colonial past. European colonialism extracted natural resources as well as people from the colonised world. In the twentieth century, international financial institutions took on the role of economic colonisers in the form of structural adjustment policies (SAPs) for the Third World. Now an ecological crisis created by the old colonisers is being reinvented as another market opportunity. This new market brings with it all the built-in inequities that other commodity markets thrive upon. From the pumping of pollution into communities of colour in Los Angeles to the land-grabbing for carbon sinks in South America, emissions trading continues this age-old colonial tradition.

Notes

1. This chapter was originally published in *Capitalism Nature Socialism*, which is hereby thanked for permission to reproduce it here. For more, see Carbon Trade Watch 2003 and TNI/FASE 2003, a joint publication of FASE-ES and Carbon Trade Watch.
2. For the purposes of this chapter, the term 'emissions trading' refers to credit-and-trade (Clean Development Mechanism and joint implementation), as well as cap-and-trade systems in the Kyoto Protocol.
3. According to the IPCC, 'There is new and stronger evidence that most of the warming observed over the last 50 years is attributable to human activities' (2001).
4. World Resources Institute website: http://www.earthtrends.wri.org/maps_spatial/maps_detail.cfm? theme=3. United States 4.6 per cent world population: http://www.earthtrends.wri.org/searchable_db/index.cfm. European Union's population grows by 1.5 million: http://www.itv.com/news/Related1428225. html.
5. The Forest Stewardship Council (FSC) is an association of environmental and social groups, the timber trade and the forestry profession, indigenous peoples' organisations, responsible corporations, community forestry groups and forest product certification organisations from around the world, which provides standards for responsible forestry (http://www.fsc.org). The FSC has been criticised by groups such as the World Rainforest Movement (WRM) for including plantations in its certification schemes. WRM argues that plantations are not forests and should not be considered for the FSC label (http://www.wrm.org.uy/bulletin/64/viewpoint. html#viewpoint).
6. Please note that DNV pointed to the lack of guidance from the official UNFCCC rules in clarifying this problem.
7. See Future Forests website: http://www.futureforests.com/.
8. Sinks in the CDM are limited to 1 per cent of Annex I countries' annual emissions. This figure is based on the average rate of growth of plantation trees. See the SinksWatch website for more information on sinks and the Kyoto Protocol: http://www.sinkswatch.org.
9. For more discussion of this point, see the WRM website: http://www.wrm.org.uy/publications/briefings/ CCC.html#sinks.
10. The six greenhouse gases focused on in the international negotiations are carbon dioxide (CO_2), methane (CH_4), nitrous oxide (N_2O), hydrofluorocarbons (HFCs), perfluorocarbons (PFCs) and sulphur hexafluoride (SF_6).
11. Interview with Frank van der Vleuten, Free Energy Europe, Netherlands office, December 2002.
12. Community Wind Turbine: Centre for Alternative Energy (CAT) website: http://www.cat.org.uk/gallery/ CWTphotodiary.tmpl?cart=32549200181239561&startat=1&subdir=gallery.
13. Renewable Energy for local benefit project, Escanda: http://www.escanda.org/.

5

World Bank Carbon Colonies

Daphne Wysham and Janet Redman

Sajida Khan lived next to the toxic dump on Bisasar Road in Durban, South Africa. She suffered two bouts of cancer before her death in July 2007 and she also lost a nephew to the disease. For more than a decade, Khan had organised in the community and launched legal challenges to get the dump decommissioned. Then came what might have been her most formidable foe: the World Bank.

The World Bank claims to be dedicated to reducing poverty and improving living conditions around the globe. Despite the fact that the dump – which was opened in this brown and black community by white rulers under apartheid – is making residents sick and despite the closure pledge made during the mid-1990s by African National Congress (ANC) leaders, the World Bank wanted to keep the dump open. Why? To capture methane, transform it into local 'clean power' and allow industrialised countries to invest in it – in exchange for continuing to pump out carbon dioxide (CO_2) back home. Investors predict that carbon could become one of the largest markets in the world. While it is open to private investors, foremost among those gearing up to profit from this new market is the World Bank.

Entering the fray

In 1997, confidential internal documents leaked to the Institute for Policy Studies revealed early plans for the World Bank's involvement in carbon trading.[1] In the same year, the US government was forging a trading scheme in conjunction with the parties to the United Nations Framework Convention on Climate Change (UNFCCC), in which emission credits were to be traded exclusively among industrial Northern countries. Brazil and other developing countries countered with the Clean Development Fund (CDF). The CDF, based upon the principle of 'the polluter pays', would have financed projects in developing countries with fines levied against Northern countries that failed to comply with the Kyoto

Protocol's emissions reduction goals. Northern negotiators transformed the CDF into the Clean Development Mechanism (CDM), which proposed market-based emissions trading between northern and southern states.

Initially received with scepticism by everyone but US government officials and one or two non-governmental organisations (NGOs), the CDM was nevertheless accepted by all parties to the Kyoto Protocol as an olive branch, to bring the heaviest polluter – the United States – on board. However, George W. Bush withdrew his country from the Kyoto Protocol and left behind the CDM for others to work out.

In a self-appointed role as broker between Northern and Southern governments and industries, the Bank planned to profit handsomely by charging a 5 per cent commission on CDM transactions, one leaked document showed. The memo noted that, with a potential market in CO_2 that could reach US$2 billion by 2005, the Bank could quickly earn US$100 million in one year. The 'commission' – which the Bank claims merely covers its costs – is closer to 13 per cent as of 2007.

None of the signatories to the UNFCCC or the Kyoto Protocol asked the World Bank to play this role. In fact, many, including US Treasury officials, actively discouraged it, recognising potential conflicts of interest. US Treasury officials wrote in an internal memorandum:

> In our viewpoint, this initiative is inadvisable for several reasons. Perhaps, most important, it would divert needed effort from reforming the Bank's mainstream power sector portfolio, which has a far greater potential impact on greenhouse gas emissions. It also places the Bank in a position of both generating and benefiting from carbon trading, which represents an inherent conflict of interest: the Bank would have little motivation for decreasing the baseline carbon emissions in its power sector projects. Finally the LDCs [less developed countries] themselves appear strongly opposed to any role for the MDBs [multilateral development banks], especially the World Bank, in the future CDM; and the proposed Carbon Fund appears to have little support among other donors. For all these reasons, we should attempt to refocus the Environmental Department and Bank management on addressing the larger, but ultimately more important, problems of the mainstream power sector portfolio.[2]

The Bank, rarely accountable to national or international governmental bodies, ignored these and other objections and simply took the task of carbon trading upon itself. The Bank worked its way into the carbon trading business in 1999 with the Prototype Carbon Fund (PCF), portraying it as an opportunity to work out the glitches in the CDM before it was launched globally and as a temporary catalyst to jump-start finance for clean energy technology. The PCF's former director, Ken Newcombe, told activists at the time that the fund would be 'entirely renewable', with solar, wind, micro-hydro and geothermal power projects making up its portfolio. As time passed, it became clear that the PCF was far from renewable; in fact, it followed the more forthright trajectory laid out in the leaked 1997 document – namely, pursuing the low-hanging fruit of the global carbon market.

By 2004, at an event jointly sponsored by the International Emissions Trading Association (IETA) and the World Bank held at the climate negotiations in Buenos Aires, Newcombe emphasised that nitrogen dioxide and hydrofluorocarbons, not renewables, were the most attractive candidates for carbon financiers, representing 1 billion tonnes in carbon dioxide equivalent (CO_2e) trade before the mechanism expires in 2012. 'One would expect that the CDM would support wind, solar and small hydro. But the CDM methodologies . . . and an un-level playing field for renewable energy . . . make it very difficult,' he lamented.[3]

A strange and bitter crop

While Sajida Khan and others worked to get the Bisasar Road dump closed, Newcombe approached the mayor of Durban in 2002, proposing that the city could profit by turning waste gas into electricity. The project would make money both by selling power locally and by reaping money from the PCF. The methane that this and other landfills produced could be siphoned off to a power plant and the city government would be rewarded with R60 million over 21 years from Northern industries reluctant to reduce their own emissions and eager to buy their way out of the problem. The Bisasar Road dump is emblematic of the problems carbon trading can lead to: allowing Northern governments to profit from carbon profligacy, while forcing the poorest and darkest-skinned in the South to pay with their lives.

In India, in June 2004, Newcombe signed a letter of intent for the purchase of 800 000 tonnes of carbon credits from the FaL-G brick and block industry at US$5 per ton of CO_2e over the next ten years. FaL-G bricks and blocks are made from fly ash, a by-product of coal combustion. Unlike clay bricks, fly ash bricks can be constructed without the use of thermal energy.

While these bricks may be more environmentally friendly from a climate perspective, there can be health hazards. 'Coal fly ash typically contains a range of heavy metals, a range of radioactive elements, a range of poly-aromatic hydrocarbons and other semi-volatile contaminants,' says Dr Pat Costner, former senior scientist at Greenpeace, noting that the people handling the fly ash are at greatest risk of exposure to these toxic contaminants.

'The thought of turning over the management of fly ash to the hundreds or perhaps thousands of tiny brick-making operations in India is enough to make the soul shiver,' Costner adds.

In Brazil, a company called Plantar owns a monoculture eucalyptus grove in the state of Minas Gerais, covering 23 100 hectares. The total land owned by Plantar, acquired by pushing local communities off their land under dictatorial regimes, is more than 120 000 hectares. The fast-growing eucalyptus trees will eventually be harvested and used as charcoal for the production of a low grade of iron. For small farmers nearby, the consequences of this plantation are devastating – streams and swamps have dried up, chemicals contaminate the air and water and the diverse species that once inhabited the land have all but vanished.

These plantations are allegedly avoiding the production of 4.3 million tonnes of CO_2 that would have been emitted had coal, rather than charcoal from the plantations, been used for smelting pig iron – that is 4.3 million carbon credits that can be sold to a Northern industry unwilling to reduce its emissions domestically by the same amount.

Is there truly a net benefit? In the next two decades, these eucalyptus trees will be cut down. The CO_2 produced by Northern industries that have bought the PCF's credits, on the other hand, will remain in the atmosphere, on average, for 50 to 200 years.[4] Elsewhere, the Bank's PCF may be used to finance the planting of genetically engineered (GE) trees that have been altered to absorb more CO_2. But, says Anne Petermann, of Global Justice Ecology Project, 'this World Bank subsidy will enable countries to develop huge GE tree plantations that destroy native forests and worsen global warming'.

The World Bank is also the only large CDM project developer promoting carbon finance for large hydropower projects. While the CDM was meant to set relatively rigorous criteria for carbon offset projects, three of the ten hydropower projects in the Bank's PCF are larger than 10 megawatts – the generally agreed standard for small hydro – and do not uphold the standards set by the World Commission on Dams, a gold standard for dam-building. In addition, deals made through the CDM are by definition supposed to provide *additional* funding for renewable energy projects, enabling the development of clean energy that would otherwise not have been financially possible. The Xiaogushan hydropower project was declared the least cost project option by the Asian Development Bank and was already under construction when the World Bank proposed supporting it with carbon credits (McCully 2005).

An expanding portfolio

Eight years after the PCF opened as a 'short-term' boost to the carbon market, it is still operational and the World Bank's role in the carbon offset industry is growing. The Bank now refers to its carbon finance as a 'mainstream' part of its overall lending programme and has added nine funds with 16 government and 66 private sector participants to its carbon finance portfolio.

In June 2004, the Bank launched the BioCarbon Fund, through which it channels money to test how land use and forestry activities can generate carbon credits.[5] Like the offset project in Minas Gerias, this fund focuses investments on commercial-scale tree plantations. In many cases, the plantations are sited on public land, or on private land traditionally used for local subsistence farming or grazing. The new monoculture plantation forests are generally deeded to private companies, or managed through murky public-private partnerships, making it nearly impossible to tell exactly who is responsible for delivering the contracted carbon credits and who gets the money from the sale of those credits. Outside Honduras' Pico Bonito National Park, families that were supposed to stay on replanted land and practise sustainable forestry, according to the Bank's plans, instead sold their parcels to the for-profit company created to run the plantation and moved to places where they could access public services. Ironically, the success of these forestry projects to reduce carbon emissions is pegged on residents relinquishing local control and giving up subsistence agroforestry traditions.

The Bank also opened a Community Development Carbon Fund (CDCF) in 2003 to 'link small-scale projects seeking carbon finance with companies, governments, foundations, and NGOs seeking to improve the livelihoods of local communities and obtain verified emission reductions'.[6] According to Bank staff, this is the only fund that requires offset projects to show clear sustainable development benefits for local host communities. As of 2007, the CDCF accounted for a mere 6 per cent of the Bank's carbon finance portfolio.

Additionally, the World Bank administers some funds for individual countries, including the Netherlands CDF, launched in 2002, the Italian Carbon Fund, launched in 2004, and the Spanish and Danish Carbon Funds, launched in 2005. A small portion of the money from these funds is pooled with private capital to create the Umbrella Carbon Fund, established in 2005, which backed a single massive project in China to reduce greenhouse gas emissions from manufacturing ozone-depleting hydrofluorocarbons. This project has been mired in controversy, with some alleging that it actually may have created perverse incentives for Chinese hydrofluorocarbon manufacturers to actually increase their production, in order to profit from the destruction of the chemical.

As of 2007, the Bank's carbon finance programme had amassed nearly US$2 billion in capital from the most polluting companies and countries in the industrialised North and channelled more than US$1 billion of that into some of the most environmentally destructive industries in the global South. Only one-fifth of the active projects are in the renewable energy sector, while more than 80 per cent of the funds disbursed have gone to coal, metal, cement and industrial gas companies. As these projects continue to grow and new funds are added to their portfolio, the World Bank is helping create – and corner – a market that undercuts its own mission. 'While the Bank's mission was to help establish a market that catalysed private sector investment in climate-friendly projects, they have now largely crowded out the private sector,' notes Ben Pearson, climate campaigner with Greenpeace Australia. 'They now dominate the carbon market.'

The Bank's fossil-fuel portfolio

The irony of the World Bank becoming a dominant player as a money-making broker in the growing carbon trade does not end there. Today, the Bank is also one of the largest public sources of funds for the fossil-fuel industry. From 1992 through late 2004, the World Bank Group approved US$11 billion in financing for 128 fossil-fuel extraction projects in 45 countries. These projects will lead to more than 43 billion tonnes of CO_2 emissions, in addition to which, more than 82 per cent of World Bank financing for oil extraction has gone to projects that export oil back to countries in the wealthy North.

The irony of this dual role – carbon trader and fossil-fuel financier – is apparently lost on the Bank, whose literature reports: 'The World Bank's carbon finance initiatives are part of the larger global effort to combat climate change, and go hand in hand with the Bank's mission to reduce poverty and improve living standards in the developing world. The threat climate change poses to long-term development and the ability of the poor to escape from poverty is of particular concern to the World Bank.' (World Bank 2004)

This state of institutional schizophrenia evolved over the last couple of decades. Since the 1980s, under pressure from the Reagan administration, the Bank has pried open developing countries' fossil-fuel sectors in order to satisfy the growing import needs of industrialised countries. In 1981, the US Treasury urged the Bank to play a leading role in the 'expansion and diversification of global energy supplies to enhance security of supplies and reduce OPEC [Organisation of Petroleum Exporting Countries] market power over oil prices'. The Treasury also noted that, as opposed to the US government, the neutral Bank could play an important role in fostering foreign corporate investment in developing countries' energy sector. The Bank implemented these directives with great success. Then came the 1992 Rio de Janeiro Earth Summit, progenitor of the Kyoto Protocol, which placed much of the financial control over sustainable development aid – and particularly clean-energy financing – within the Bank's confines.[7]

Research conducted by the Sustainable Energy and Economy Network (SEEN), founded in 1996 at the Institute for Policy Studies, tracked how well the Bank was holding up its end of the bargain. Among other problems, the last decade has seen unprecedented levels of Bank financing for fossil-fuel projects, especially those that export oil to Northern markets and threadbare support for renewable energy and energy-efficiency projects. In an average year, the World Bank supported fossil-fuel projects with lifetime emissions of 1 457 megatonnes of carbon. This figure is at least 4 – and as much as 29 – times the amount of annual emissions reductions anticipated under the CDM.

The primary direct beneficiaries of these fossil-fuel projects are Northern corporations, particularly those based in the United States. They benefit either through direct loans or through the privatisation process enforced by Bank loans. Halliburton and Enron, to name two such primary beneficiaries, enjoyed global expansion in the 1990s hand-in-glove with World Bank Group project financiers (SEEN 2007a; 2007b). And the Bank's impact reaches far beyond the projects it funds. Its decisions potentially affect more than 80 per cent of all private banks – those so-called Equator Principle banks that base their standards upon those of the World Bank – and all the public banks that also look to the venerable institution for guidance.

Refusing to reform

Over the past fifteen years, many people – from the world's most disenfranchised to Nobel laureates and internal whistleblowers – have tried to convince the Bank to realign its energy portfolio, but the Bank has ignored their pleas. A brief look at the history of the World Bank's approach to energy development provides an important perspective on the institution's current attempts to position itself at the forefront of the effort to combat climate change.

NGOs began confronting the World Bank's skewed energy lending in 1992. After a lengthy consultation process with NGOs and endorsement by the Bank's board, the Bank produced two policy documents on energy issues, such as 'The World Bank's Role in the Electric Power Sector' (World Bank 1993), which laid out the following policies:

- a commitment to transparency in decision-making;
- an agreement that least cost energy planning, long pushed by environmentalists, should be advanced;
- an agreement that energy subsidies for fossil fuels and other traditionally environmentally unsustainable energy resources should be removed;
- an agreement with environmentalists that demand-side management and energy efficiency were approaches to be preferred over energy expansion;
- 'pollution reducing technology' needed to be more aggressively pursued in its energy lending; and
- concurrence with environmentalists that all of the above policies should be integrated into dialogues with its clients and given high visibility in loan agreements.

These principles were praised by environmentalists, but, like many preceding commitments, have proven hollow over time; they have not been backed by the financial commitments necessary to make them a priority within the Bank.[8] In August 1996, the Bank went back on its own binding operational policies on energy, downgrading them to non-binding good practices. Had these policies remained in force, they alone could have had a significant impact on the global climate. The Bank's legal department argued that because of procedural issues, the operational policies were not bona fide policies that could be enforced within the institution.

The Bank made a foray into solar energy that also fell on barren ground. The Bank's Solar Initiative, launched in 1994, is a good example. The Initiative aimed to raise awareness among Bank staff and clients about how to commercialise renewable energy technologies. However, the budget for these technologies was insignificant and the Bank demanded that renewables be promoted without subsidies – a tactic that fails to consider the unequal playing field between renewable energy and more established non-renewable energy resources.

A 1995 Bank 'carbon backcasting study' also proved an exercise in futility. The study, by the International Institute for Energy Conservation, Hagler Bailly, and the Stockholm Environment Institute (1997), looked at how the Bank's energy portfolio would have been affected if carbon emissions had been considered in project selection. The study found that if the cost of expected carbon emissions had been calculated at US$20 per ton, renewables would have become more attractive investments and coal a pariah. Although this exercise engaged a variety of expert scientists and economists, it resulted in no changes in World Bank energy policy.

Even though the Bank houses the Global Environmental Facility (GEF), in 1998 the GEF bit the hand that feeds it, echoing civil society concerns regarding investments in fossil fuels and climate change:

> The Bank has not succeeded in systematically integrating global environmental objectives into economic and sector work or into the CAS [country assistance strategy] process; nor has it taken meaningful action to reduce its traditional role as financier of fossil fuel power development . . . Continued financing by the World Bank for such projects (as conventional fossil fuel generation) is inconsistent with mainstreaming of the global environment in the Bank's regular operations. (GEF Secretariat 1998)

Perhaps in response to all of these criticisms and calls for accountability, the World Bank then developed a document entitled: 'Fuel for Thought: An Environmental Strategy for the

Energy Sector'. First issued in 1998, then revised and reissued in 1999, this report again disappointed environmentalists by failing to deliver on a promise to provide targets and timetables for renewable energy lending, or to address the need for energy poverty alleviation. The final report contained no clear targets for renewable lending, nor did it embrace proposed methodologies to account for expected greenhouse gas emissions from each project before its approval. The Bank also ignored suggestions to put in place clear energy-efficiency screening requirements to improve the design of projects. Furthermore, the final report did not emphasise meeting the energy needs of the two billion rural poor who lack basic energy services.[9]

Finally, the Extractive Industries Review (EIR) represented the peak of Bank arrogance. With global civil society clamouring for change, the former Bank President James Wolfensohn launched the EIR in 2001, to determine the effect of extractive industries on the world's poorest people. This exhaustive study, involving government, environmentalists and business, was presented to the board in 2004.

The report recommended that the Bank

> phase out investments in oil production by 2008 and devote its scarce resources to investments in renewable energy resource development, emissions-reducing projects, clean energy technology, energy efficiency and conservation, and other efforts that delink energy use from greenhouse gas emissions. During this phasing out period, WBG [World Bank Group] investments in oil should be exceptional, limited only to poor countries with few alternatives.

Among the report's other recommendations were that the Bank should continue its moratorium on lending for coal and increase lending for renewable energy by 20 per cent annually.[10]

After lobbying by key staff, the Bank's board of directors rejected most of the report's recommendations, but chose to implement a target of a 20 per cent increase each year for five years in renewable energy lending and greater transparency in terms of oil industry revenues. In the fiscal year 2005, the Bank barely achieved this 20 per cent increase, relying heavily on a rise in carbon finance and GEF projects to demonstrate the increase.

G8 power

The Group of Eight (G8) industrial nations are the power behind the Bank.[11] In 2001 they met in Genoa, Italy and discussed a proposal that would commit rich nations to help one billion people around the world get to their power from renewable energy sources. Among the final report's recommendations was a proposal to encourage a shift in the priorities of international lending agencies, like the World Bank, to support more clean energy projects in poor countries. The Bush administration killed this initiative, however, which had the support of the much of the rest of the G8.

After all of these failed attempts at meaningful change, at the close of the 2005 G8 Summit in Gleneagles, Scotland, G8 leaders requested the World Bank to work on creating a new investment framework on climate change. In response, the Bank drafted an agreement on 'long-term climate management beyond 2012', after the first commitment period of the Kyoto Protocol ends. The climate language has since been taken out of the title, at the behest of Bush administration officials and the current proposal is called the 'Clean Energy and Development: Towards an Investment Framework'.

The World Bank's functional schizophrenia

By the twenty-first century, the World Bank had become the leading fossil-fuel financier and the number-one broker of carbon trades. Now, it is adding another, even more schizophrenic role to its identity: climate change investment adviser to the world.

'The Gleneagles Dialogue', according to the organisation's UK press office, 'has been designed to create a space away from the formal negotiating process at the United Nations (UN) to discuss new ideas, identify common ground and practical actions to reduce greenhouse gas emissions'. Because it effectively brings together the world's wealthiest countries with the world's most populous countries to discuss ways of moving forward on climate change, the Gleneagles Dialogue is not an idle exercise. One would have thought, based on the understanding that climate change poses a particular risk to the poorest in the developing world, that such an investment framework pulled together by the World Bank would set as its goals targets that included the most optimistic scenarios put forward by the International Energy Agency (IEA), at the very least, to avoid greater harm to the poorest.

Instead, what the World Bank penned in the paper entitled, 'Clean Energy and Development: Towards an Investment Framework', approved in April 2006 by the World Bank and the International Monetary Fund's (IMF's) development committee and reissued in August 2006, is a prescription for planetary disaster: business as usual, a sea-level rise of at least three feet over the next century and the massive extinction of a large share of the world's plant and animal species. Of course, among those most threatened are the poorest people of the world.

Catastrophic targets

The Bank admits in its framework that if atmospheric CO_2 increases above 450 parts per million, a global temperature rise beyond two degrees Celsius above preindustrial levels is almost assured. The two-degree limit has been adopted by the European Union and NGOs as the threshold for dangerous climate change, beyond which further temperature rise would cause 'massive species extinctions and dramatic changes in ecosystems [which] will have severe consequences for human well-being' (IUCN 2005). However, the Bank's framework considers scenarios that range from 450 to 1000 parts per million, sending a message that it might be a viable option to allow carbon levels to reach well beyond 550 parts per million – despite the fact that some scientists believe that even at 400 parts per million, there is up to a 26 per cent risk of overshooting the two degrees Celsius target.

In addition, the World Bank draft assumes a 60 per cent growth in greenhouse gas emissions by 2030 as its 'reference case' scenario, something that far surpasses the goals of the IEA alternative scenario. The reference scenario would see hundreds of millions of people in less developed countries put at higher risk of starvation and disease, or turned into environmental refugees due to a sea-level rise that could reach three feet or more within a century. Recent scientific reports suggest such an increase would place between 1.2 billion and 3 billion people at risk of water shortages, and would cause millions of tonnes of cereal crops to fail. The IEA's alternative scenario would result in a 30 per cent increase in energy-related CO_2 emissions by 2030 above 2002 levels. However, even a 30 per cent increase in energy-related emissions is unacceptable.[12] For the world's poor, such scenarios could mean massive drops in agricultural productivity, loss of access to essential natural resources and dramatic declines in living standards. Therefore, it is critical that the Bank include in its scenarios other scenarios, including those surpassing the IEA's scenarios, which were not intended to place a limit on what is possible, but merely to illustrate what greenhouse gas emissions could be achieved at the lowest possible cost. With the future of the planet at stake, cost should be our final – not our first – consideration.

The Bank has chosen a far less ambitious approach than that set by the G8's renewable energy task force in 2001, which imagined targeting one billion people with renewable energy by 2010. But there are many other profound flaws in the World Bank's latest paper.

No climate footprint for the Bank

The Bank does not currently calculate the climate footprint of its own investments. According to studies conducted by the Institute for Policy Studies, the Bank's investment in fossil fuels over the next twelve years will release the equivalent of two years' worth of global greenhouse gas emissions. Yet nowhere in the World Bank's books are calculations of its climate impact considered. Instead, the Bank claims that it has reduced 'the average intensity of carbon dioxide emissions from energy production' and 'average energy consumption per unit of GDP', despite never having made such an assessment publicly available.

Because the World Bank does not determine its own climate footprint, it has led the entire financial community (including private financial institutions that have signed on to the so-called Equator Principles and follow the World Bank's lead on environmental guidelines) down a 'see-no-evil' path of continued fossil-fuel finance, claiming credit for carbon emissions captured, but never transparently revealing their role in carbon emissions released.

Bank strategies for energy and climate change

The Bank approach to energy delivery follows the primary recommendations in the G8 investment framework: privatise and deregulate energy markets. There are enormous flaws in this model that have resulted in corruption (such as at Enron), human rights abuses and ultimately, an increase in greenhouse gas emissions. For example, the Chad-Cameroon pipeline would not have been built but for World Bank energy finance; yet the greenhouse gases that will be released are not counted as a debit in the World Bank's statement of carbon accounts.

The Bank also promotes nuclear power as a solution to climate change, despite massive subsidies that are its only key to viability and global concern over nuclear accidents and terrorism (for more information, see Bank Information Center et al. 2006: 6). The Bank acknowledges that hydropower will become unpredictable as rainfall patterns change, but it still recommends this technology as a solution to climate change, in spite of a World Commission on Dams finding that in tropical areas, plant decay releases enough methane to push greenhouse gas emission rates for hydropower above those of coal-fired power generation (Bank Information Center et al. 2006: 18).

In terms of fossil fuels, the World Bank promotes integrated gasification combine cycle (IGCC) and carbon capture and storage, both untested technologies for use on coal-fired power plants, despite the problems associated with coal burning, to the detriment of renewable energy, such as wind, a fast-growing, competitive and clean source of energy. But the IGCC uses synthetic gas as its main fuel. The synthetic gas is produced when coal is heated up, but not burned, and gas is given off – also known as 'gasification'. The IGCC process uses steam generated as waste heat to drive another turbine, thereby the name 'combined cycle'. 'Clean coal' is a term given to coal that has been chemically washed of minerals and other impurities. It is then burned, perhaps in an IGCC power plant and there turned into a gas, with the sulphur dioxide and nitrogen oxide siphoned off and the CO_2 captured in the gasification process and stored underground, instead of being released into the atmosphere.

While it sounds promising on paper, there are no proven CO_2 storage facilities. Storage of the CO_2 is envisaged either in deep geological formations, deep oceans, or in the form of mineral carbonates. The international community is currently exploring long-term storage of CO_2 in sub-seabed geological formations, including old oil and gas wells and saline aquifers, as part of a suite of options for climate change mitigation. Yet recent studies show that CO_2 stored underground is causing a chemical reaction that may end up dissolving the very mineral that helps keep the gas underground.

While carbon capture and storage advocates claim that it could reduce greenhouse gas emissions from coal and other fossil fuels by 80–90 per cent, they often fail to mention how much energy the process of carbon capture and storage requires – anywhere from 10–40 per cent more than normal. This could increase both the price of the plant by 30 to 60 per cent and the quantity of fuels needed to capture the carbon. This, in turn, has environmental impacts – both at the source of mining for fossil fuels and when the waste products, such as mercury – a common by-product of coal burning – are disposed of.

After capture, the CO_2 must be transported to suitable storage sites. This is done by pipeline, which is generally the cheapest form of transport or by ship when no pipelines are available. Both methods are currently used for transporting CO_2 for other applications. The energy costs associated with transportation are not insignificant.

Ironically, methane, the desired by-product of IGCC – the fuel that is burned – is being flared in the Niger Delta and other regions where oil is extracted. Methane is also a by-product of waste decay. Biogas digesters, which can use human or animal waste, can produce methane to be burned with no hazardous mining or toxic by-products such as mercury.

The Bank in Bali

What was gained at the 2007 UNFCCC climate negotiations in Bali, where the Bank held sway? After two weeks of climate talks that brought together nearly 190 countries and more than 10 000 delegates, observers and activists, there was very little to show for negotiations that were less about urgent climate action than business as usual. The Bali plan to guide the transition from the Kyoto Protocol to a new post-2012 agreement entrenches the power of big business and global financial institutions, like the World Bank, that work on its behalf, without committing any government to concrete emissions reductions.

The proposals put forward in the Bali agreement for reducing greenhouse gas emissions and adapting to climate change lack concrete detail (all references to binding targets were removed). But they are spelled out clearly enough to see that trading in carbon credits – both through existing Kyoto Protocol mechanisms and newly envisioned sectoral schemes – will likely be at the centre of the next global treaty. Outlined in the roadmap is an adaptation fund that could reach US$500 million by 2012, which would be administered by the GEF, with the World Bank acting as financial trustee, notwithstanding strong reservations by 'most' developing countries, according to one official announcement. One funding proposal suggests bolstering donations to the fund from industrialised countries by recouping a 2 per cent fee on revenues from carbon offset projects carried out under the Kyoto Protocol's CDM.

Proponents of the adaptation fund claim that by using the CDM, rich countries would be 'forced' to finance clean energy projects in poorer countries. But the fund's total capital is almost insignificant compared to the US$50 billion that Oxfam estimates the developing world will need every year to cope with climate changes. However, by naming the CDM as a major source of funding for adaptation, the Bali mandate entrenches carbon trading and the World Bank in future negotiations. The proposal ensures that developing countries, eager for a way to pay for responses to expected climate disasters, have an increasingly vested interest in seeing market mechanisms flourish. And as the institution that both promotes new clean development methods and brokers offset finance for emerging technologies, the World Bank has an increasing stake in the carbon market, as well.

One of the most controversial debates that emerged from the Bali negotiations was the role of forests in combating climate change. Considering the Bank's track record in the BioCarbon Fund, it was surprising to see the international community call on the Bank to take the lead on reducing emissions from deforestation in developing countries, a process known by the acronym REDD (Reducing Emissions from Deforestation and Degradation). By including deforestation in the post-2012 roadmap, standing forests essentially get folded into the carbon market. But the Bali plan does little to explain how forested countries and the communities who depend on forests for their survival, would be compensated for slowing deforestation. And the Bank didn't bother consulting with forest communities and indigenous peoples in setting up this plan, to the consternation and protest of civil society groups gathered in Bali.

The World Bank has stepped into this vacuum to guide a market in REDD credits through its newly launched Forest Carbon Partnership Facility (FCPF). The fund will select countries to try a new approach to carbon trading by setting national emissions reduction targets for a country's entire forest sector, instead of creating baselines and targets on a project-by-project basis as with the CDM.

In response to protests from indigenous peoples and sustainable forestry groups, the Bank is pledging three additional months of global consultations with stakeholders. But those people who will be most affected by the Bank's new initiative point out that programme elements are already locked in place, making this consultation moot. Like the BioCarbon Fund, there is nothing built into the facility to ensure that the benefits of a global forest trading scheme would reach forest peoples. Bank staff have been non-committal in explaining where social and environmental safeguard policies that offer recourse to affected communities would apply. And because of the programme's large scale, monitoring local impacts will be more difficult than under CDM projects. Worryingly, there are no provisions in the FCPF to prevent international logging operations from profiting from forest carbon credits. Critics familiar with the World Bank's current forestry programme have raised warnings of massive displacement as companies rush to acquire forested land and governments shift public policy to facilitate industrial land-grabs. The programme also unfairly rewards those countries that have deforested their lands most rapidly in recent years, potentially encouraging other countries to follow suit.

Investors, however, are quite pleased with the idea, having long asked the Bank to establish consistency throughout the carbon market. Under the new World Bank initiative, private investment companies would have an easier time assessing the risks of putting money into the carbon market and would lower their transaction costs by purchasing credits from a large number of carbon offset forestry projects at the same time. Aside from its work dealing with forests, the Bank has already begun work on a new carbon partnership facility to expand the sectoral approach into markets for carbon credits generated from power sector development, gas flaring, energy efficiency, transportation and waste management systems in developing nations.

The growing role of the World Bank in clearing a path for private capital in an expanded carbon market was not lost on climate justice groups in Bali. Close to one hundred activists from around the world demonstrated outside the conference room where Bank President Robert Zoellick led inaugural ceremonies for the FCPF. World Bank side events and press conferences were peppered with demands for the Bank to get out of the carbon market and activists raised the larger question of whether the world can continue to afford the World Bank.

Fix or nix the World Bank

The World Bank Group has had ample opportunity to prove it could lead the global energy sector into a more sustainable, renewable and equitable future. Instead, it has become an enforcer of the status quo on behalf of the world's most powerful countries and corporations. Its energy programmes have utterly failed to curb climate change and alleviate poverty.

Those who embrace the Bank as an impartial and honest carbon broker ought to be aware that this institution's investments are driven in large part by oil-hungry nations. Unless the Bank's power structure is rewired, it will remain beholden to the world's most powerful polluters. Among the solutions being put forward by activists concerned about the lack of accountability at the World Bank are more input from elected officials on Bank policies and projects, as well as more input by the UN. Currently, the Bank remains relatively independent of both.

Those aiming to fix the World Bank have, on various occasions, proposed the following kinds of reforms:

- uphold all of the recommendations contained in the EIR submitted to the World Bank in 2004;
- openly calculate greenhouse gas emissions that will be released as a consequence of all World Bank lending before project approval, with transparent guidance for this methodology provided by the Intergovernmental Panel on Climate Change (IPCC);
- set an immediate benchmark for reduction of greenhouse gas emissions associated with projects for which the World Bank provides financing of 20 per cent per year;
- include legally binding language to restore areas degraded by oil, gas and coal development by the corporations or public entities that are responsible;
- make public, as part of the country assistance strategy (CAS), an integrated energy strategy. Each CAS should establish specific goals for improving the productivity of energy use targeted at the poorest and developing renewables and energy efficiency projects;
- in consultation with environmentalists, conduct a formal and transparent evaluation of the success or failure of the World Bank's energy lending in reaching the two billion rural poor who are without access to energy for human needs – for cooking, heating and lighting – as well as the success or failure in providing for the transportation needs of the poorest. This assessment should then be used to provide an approach to better meet the energy needs of the rural poor.

These are feasible but unlikely reforms, given the orientation of the Bank under neo-conservative rule (former Iraq war architect Paul Wolfowitz ran the Bank from 2005–07 and was followed by former US trade representative Robert Zoellick, also a member of the ultra-right Project for a New American Century). A more radical approach to diminishing World Bank damage is being taken by a growing body of civil society in the global South and North – now spanning five continents – that believes the institution is fundamentally non-reformable: a boycott of the World Bank to see the institution shrunk and eventually abolished. In Northern countries, the boycott goes after the World Bank's bond financing; in Southern countries, the movements actively reject the institution's presence in their communities, social sectors and economies.

In the meantime, most activists can agree that there is a need to get the World Bank out of carbon trading and to end World Bank lending for fossil fuels. Sajida Khan's case is inspiring, as her complaints in the environmental impact assessment (EIA) process at the Bisasar Road dump apparently forced the Bank to back off, fearing ongoing bad publicity. Her death in 2007 has given us an opportunity to reflect and expand upon this partial victory.

Finally, as CDM gimmicks and the carbon market itself start to fall apart, it may be time to dust off that proposal the Brazilians tabled back in 1997 in Kyoto. 'Polluters pay' is a better strategy than the World Bank's approach: 'polluters (and logging companies) profit'.

Notes

1. See http://www.seen.org/pages/ifis/wbstill/wbgrafx.shtml and 'How the World Bank's Investment Framework Sells the Climate and Poor People Short', September 2006, http://www.foe.org/new/releases/september2006/worldbank9172006.html.
2. The internal memo, leaked to the authors, is entitled 'International Financial Institutions and Climate Change' (draft 17 February 1998).
3. Presentation to COP 10 participants, summed up by Jim Vallette, Buenos Aires, 11 December 2004, http://climatejustice. blogspot.com/2004/12/special-report-from-inside-world-bank.html.
4. See CDMWatch and SinksWatch (2004). See also http://www.climnet.org/pubs/CANEuropePlantar.pdf and http://www.cdmwatch.org/controversy.php.
5. World Bank Carbon Finance website: http://carbonfinance.org/biocarbon/home.cfm.
6. World Bank Carbon Finance website: http://carbonfinance.org/cdcf/router.cfm? Page=About.
7. As a news report summarised, 'The developing nations had also been rebuffed in their attempt to establish a "Green Fund" under Third World control for the distribution of environmental aid. Officials from the Group of Seven leading industrial nations June 4 [1992] had affirmed that they would insist on channelling ecological aid through the Global Environment Facility, a branch of the World Bank' ('Earth Summit Held in Brazil', 1992).
8. An Environmental Defence Fund and Natural Resources Defence Council report in 1994, 'Power Failure', concluded that only 2 out of 46 electricity loans were consistent with the Bank's own policies. A WWF study, commissioned in 1996, examined 56 energy loans and found only 3 that complied with the policies that were endorsed by the Bank's board in October 1992.
9. The country assistance strategy (CAS) describes the Bank's priorities and planned lending and non-lending activities in a borrowing country over a period of three to five years.
10. Some of the other key recommendations of the EIR to the Bank include: Adopt free, prior and informed consent so that affected communities and indigenous populations have a voice in development and decision-making; recognise and adopt human rights and core labour standards; recognise 'no-go' zones for biologically and sociologically diverse areas and avoid funding projects in them and require transparency in revenue flows to companies, governments and communities.
11. Participating countries in the G8 are: Canada, France, Germany, Italy, Japan, Russia, the United Kingdom and the United States. In the G+5 are: Brazil, China, India, Mexico and South Africa. Other participants include: Australia, Indonesia, Iran, Nigeria, Poland, South Korea, Spain and the European Commission.
12. Many scientists, including the chief scientist at the National Aeronautics and Space Administration (NASA), Jim Hansen, conclude that neither scenario comes close to getting us where we need to go to avoid dangerous climate change. The business-as-usual scenario could result in a sea level rise of up to 80 feet and the massive extinction of 50–90 per cent of the world's plant and animal species. The 'reference' scenario would result in a sea level rise of about three feet by the end of this century, the extinction of a vast array of species and the drying up of water supplies in East Africa and South Asia.

6

Prototype Carbon Fund Beneficiaries

Larry Lohmann, Jutta Kill, Graham Erion and Michael K. Dorsey

Who is taking the greatest interest in carbon markets and who is best positioned to benefit from them – those who are likely to lead a transition away from fossil fuels, or those with the greatest incentives to delay that transition?

Major oil corporations such as BP and Shell, both enthusiastic initiators of internal emissions trading schemes, have never voiced any serious intention to curb their main activities of oil exploration and production. Although it has changed its name to 'Beyond Petroleum', for example, in 2002 BP committed itself to expanding its oil and gas output by 5.5 per cent per year over the next five years. Its emissions in 2001 were equivalent to almost two years of carbon dioxide (CO_2) emissions from the United Kingdom (ENDS 2002b: 4). The firm's investment in renewable energy remains at 1 per cent of the US$8 billion that it spends on fossil-fuel exploration and production every year (McGarr 2005).

Similarly, the World Bank, a determined supporter of greenhouse gas trading through the Kyoto Protocol, has scorned the August 2004 recommendation of its own review commission that it halt support for coal extraction projects immediately and phase out support for oil extraction projects by 2008 (Vallette, Wysham and Martinez 2004: 2). The commission, chaired by former Indonesian environment minister Emil Salim, pointed out that such extractive projects did nothing to promote the Bank's stated mission of alleviating global poverty. Instead, the Bank treats its carbon trading wing as what one prominent former staff member, in a personal communication with the authors, scathingly refers to as an 'epicycle' of an overwhelmingly fossil-oriented approach to energy and transport.

This approach follows the 1981 demand of the US Treasury that the World Bank play a leading role in the 'expansion and diversification of global energy supplies to enhance security of supplies and reduce OPEC [Organisation of Petroleum Exporting Countries] market power over oil prices' (Vallette, Wysham and Martinez 2004: 5). The World Bank remains one of the largest sources of public funds for the fossil-fuel industry. In an average year, the Bank supported fossil-fuel projects with lifetime emissions of 1 457 billion tonnes of carbon – a figure '4–29 times the amount of emissions reductions anticipated under the CDM per year' (Wysham 2005).

From 1992 through to late 2004, the World Bank Group approved US$11 billion in financing for 128 fossil-fuel extraction projects in 45 countries – projects that will ultimately lead to more than 43 billion tonnes of CO_2 emissions. This is hundreds of times more than the emissions reduction that signatories to the Kyoto Protocol are required to make between 1990 and 2012. Another US$17 billion has gone into other fossil-fuel related projects. More than 82 per cent of World Bank financing for oil extraction has gone to projects that export oil back to wealthy Northern countries. Bank financing for fossil fuels outpaces renewable energy financing by seventeen to one (Vallette, Wysham and Martinez 2004: 3).

Many corporate investors in the Prototype Carbon Fund (PCF) – the Bank's flagship carbon fund, set up to facilitate projects that allegedly reduce greenhouse gas emissions – are in fact receiving far greater amounts of Bank financing for fossil-fuel projects that produce emissions.

The involvement of BP and Statoil in the PCF is particularly notable given the ongoing financial support by the Bank's International Finance Corporation (IFC) for their efforts to open up the massive Azerbaijan oil fields for consumption in Western Europe and the United States. In October 2003, BP and Statoil were part of a group of corporations who received US$120 million from the IFC for development of the Azeri-Chirag-Guneshli oil fields in Azerbaijan. Greenhouse gas emissions from the oil produced by this project will be over 2 000 million tonnes of CO_2 over twenty years.

In November 2003, the IFC approved another US$125 million for the Baku-Ceyhan pipeline between Azerbaijan and Turkey, for an investment consortium that is again led by BP. An estimated 3 billion tonnes of CO_2 will be released to the atmosphere through the burning of the oil that will be transported by the pipeline.[1] Similarly, just five months after the PCF was launched in 2000, the Bank approved over US$551 million[2] in financing for the Chad-Cameroon oil pipeline. The financing package for the pipeline came to about three times the capitalisation of the PCF and the expected lifetime emissions of approximately 446 million tonnes of CO_2 (EMS 2003) generated by the pipeline's oil amount to roughly three times the 142 million tonnes that will allegedly be saved by PCF projects in total.[3]

Significantly, PCF investors get carbon *credits* from PCF projects, but no *debits* for their Bank-supported projects involving fossil-fuel extraction or use.

Who wins from World Bank PCF funding?

Corporation	PCF contribution (US$ million) for CDM and JI projects 1999–2004[4]	Received from World Bank (US$ million) for fossil-fuel projects 1992–2002[5]
Mitsui (PCF and Bio-Carbon Fund)[6]	16	1 807.5
BP	5	938.8
Mitsubishi	5	403.6
Deutsche Bank	5	165.6
Gaz de France	5	138.9
RWE	5	138.9
Statoil	5	242.3
Total	46	3 834 600.0

Source: Ben Pearson

Notes

1. See http://www.seen.org.
2. See http://eireview. info/doc/EOanalysis0209FINAL.doc.
3. Based on figures provided in the PCF's 2004 annual report (http://www.prototype carbonfund.org). Because some of the PCF's projects would have happened without the PCF and thus cannot represent real reductions, the word 'allegedly' is necessary here.

7

Big Oil and Africans

groundWork

Oil is not only unsustainable because it is so polluting and because it will run out, but also because it offers an energy future principally for the elite. Ordinary people living next to oil wells and oil refineries are most certainly not getting the benefits promised. In fact, they are the scapegoat for the elite, for they not only have to live with the pollution of oil production processes, but they also do not have access to the very energy for which they are made to sacrifice their health and wealth.

Africa's oil rush

The Gulf War has pushed up the price of oil and reinforced anxieties about security of access to crude supplies both for countries and corporations. It has added impetus to Africa's oil rush, but did not initiate it. Corporations have always been anxious to be in on the next big thing, lest they should find themselves excluded later. This is particularly so as the number and size of new discoveries globally is falling while demand is rising. The Gulf of Guinea off West and Central Africa is viewed by the oil industry as the world's premier hotspot, soon to become the leading offshore oil production centre.

Security and the cost of crude supplies are also at the top of the agenda for consuming countries. The United States, in particular, has stepped up diplomatic and military activity in the region, edging in on the regional hegemonies of the former colonial powers of Britain and France.

The international financial institutions – the International Monetary Fund (IMF) and the World Bank – are key actors in support of the Northern agenda. The World Bank itself has a direct financial interest in oil and gas. The bulk of lending by its private sector financing arm, the International Finance Corporation (IFC), is for resource extraction and the IFC makes its best profits from these loans. The Bank also acts to leverage capital from private financial institutions, which are, of course, concerned with the profits of oil debt. Its presence as a lender provides political cover. It reassures both oil and finance corporations that they will get their profits out of projects in unstable countries. Thus, the financial arrangements for the Chad-Cameroon pipeline ensure that the interest owed by these countries is paid before they see the money.

Producing countries and would-be producers are no less enthusiastic. Their economic interest is primarily in oil revenues and balance of payments, although much lip-service is also paid to technology and skills transfer. Most have expanded production to cash in on current high prices, while new exploration concessions have been awarded in almost all African countries, even where the hopes of finding oil seem slim. Everyone, it seems, is doing well by the escalation of prices – except ordinary people in oil-producing countries. While the fabulous wealth of oil is paraded before them, they have been driven ever deeper into poverty. The very common association of oil wealth with the impoverishment of people and the failure of national economies has given rise to the notion of the resource curse.

The biggest producer, Nigeria, was also the first producer in sub-Saharan Africa. Nigeria and Gabon were already major producers by the time of the first oil shock and both joined the Organisation of Petroleum Exporting Countries (OPEC) in the early 1970s and established national oil companies as part of their assertion of national sovereignty rights. At the time of the second oil shock in 1979, Nigeria felt confident enough to nationalise BP's holdings on the grounds that it was breaking the oil embargo against apartheid South Africa. BP's assets were turned over to the Nigerian National Petroleum Company (NNPC), giving it a 50 per cent holding in the Nigerian industry. Both countries also learnt to drive better bargains with the corporations. However, Nigeria was racking up debts on the security of oil during the 1970s. It was thus exposed to the debt trap when commodity prices collapsed in the 1980s and it was one of the first OPEC nations to break ranks on oil prices, as Northern powers reasserted their grip on producers.

These global scale manipulations were replicated in the activities of oil corporations on a national scale. The best evidence for this came to light in a French trial that resulted in the conviction of 30 senior Elf executives in 2003 for defrauding the corporation. Elf was the largest corporate producer in sub-Saharan Africa, with a dominant position in Francophone countries and major interests in Nigeria and Angola. The Elf system revealed at the trial was described in detail by Global Witness in 2004. It involved the systematic corruption of African leaders through a variety of kick-backs, under-invoicing on crude oil bought from Elf's subsidiaries to skim the revenues owed to African countries and the peddling of oil-backed debts, with the specific intention of creating a perpetual dependency on Elf. The debt system was purposely obscure, so that Africans were only aware of the official lending bank, while Elf itself profited from the debt. It also profited from facilitating arms deals financed by the debt.

Elf's activities in Nigeria during the 1990s are also under investigation. But it is certainly not the only corporation to have instigated corruption. The French investigations have in turn led to investigations into allegations that the US corporation Halliburton was implicated in bribery in Nigeria. Allegations of corrupt dealing, price manipulation, political string-pulling and complicity with state brutality have also haunted Shell's operations in Nigeria.

Working for the United States

While Chevron has been active in Africa for some time, the US corporations are prominent in the new oil fields – offshore of Nigeria, as well as in the new petro-states. Nigeria's take of oil revenues contrasts with the very poor deals done by latecomers. Equatorial Guinea gets 10 to 20 per cent and Chad only 10 per cent. The oil project in Chad, however, would not have gone ahead but for World Bank participation because of the level of 'political risk'. Chad initially negotiated its deal in 1988 and subsequently tried to revise it in 2004. Despite approximately US$1.6 million in World Bank-financed legal assistance, the Chadian government was able to negotiate only a marginally better deal in the new convention (Gary and Reisch 2005: 39).

The World Bank justified its participation in the project on the grounds that there was no other developmental option in Chad, that its participation would ensure that Chad would escape the resource curse and the project would thus contribute to poverty alleviation and that poor Chadians need access to modern energy. Critics argued that Chad's governance and human rights record made it a dead-ringer for the resource curse and that a project focused entirely on exports was scarcely conceived to access energy for poor people. Thus far, the experience of the project confirms the critics' view.

Producing environmental injustice

The industry likes to talk of the production chain as a value chain. The value added at each link in the chain includes salaries and wages, taxes and other payments to governments, debt repayments and interest and, finally, what is taken as profits by the corporations and either paid out to shareholders or reinvested. It excludes the costs of raw materials and services provided by other businesses. Value added is thus the difference in value between what comes in and what goes out.

The notion of value added serves a vital ideological function. It proclaims that what it counts as value amounts to a general social good and this proclamation is then turned into an assumption. Thus, a country's gross national product (GNP) is, put simply, the aggregate of value added from all economic activity. The basic assumption of mainstream economic thought is that the growth of GNP is in everybody's best interests, even if some people benefit more than others.

Yet the notion of value added conceals more than it reveals. The calculation of value excludes major costs, which are also produced at each link in the chain and imposed on other people, on society in general, or on the environment. These costs could be called value subtracted, although they are more conventionally known as externalities. Those who pay these costs – those from whom this value is subtracted – are those who are made poor by the process of wealth creation.

Enclosure

Enclosure involves the appropriation of a common resource and the dispossession of those who previously had rights to the resource.

Nigeria's Land Use Act, promulgated in 1978 by the then military regime under President Obasanjo, gives the state control of all land and allows it to evict people where land is required in the 'over-riding public interest'. The public interest specifically includes the requirement of the land for mining purposes or oil pipelines. The Petroleum Act makes oil and natural gas the property of the federal state. It provides for compensation for loss of use, but any rent on the land goes to the state.

The practical effect of these two acts is that oil corporations can and do take what they want from the people within their areas of operation. The corporations themselves call this the 'land take'. They also decide what they will pay in compensation. The land take is enforced by the state security forces.

The elite of the Mobile Police is deployed within the oil installations and paid by the corporations at well-above normal rates. They are commonly known as the 'Shell police' or the 'Chevron police' etc. On at least one occasion, Shell made a deal to supply arms to security forces. It denied doing so until confronted with evidence and then claimed that the deal had fallen through. Various gangs recruited from the ranks of unemployed youth and armed with anything from machetes to submachine guns have also been deployed to intimidate and terrorise people.

This is the pattern for the oil industry throughout Africa: the corporations are given the right to take what they want, while all rents, royalties and other monies received in exchange for this right are taken by the state. Consultation with communities in Chad is touted by the World Bank as a model of best practice. Consultation, however, comes after the negotiation between the state and the corporations has already expunged people's rights to the land and made them over to the corporations. That the oil project will go ahead and that the land required by the project will be appropriated is not up for negotiation in the course of consultation. Even the parameters of compensation are predefined. What is left to consultation amounts to little more than a public relations exercise.

Externalisation

Externalisation is about excluding the costs of pollution from the value chain, so that these costs do not appear in the market price of the commodity. Externalised costs are thus made to constitute free benefits to the corporate producer. They are an unacknowledged subsidy, but these costs do not in fact disappear. Rather, they are imposed on others: they reappear as uncompensated costs to communities and workers who suffer the loss of resources and health damaged by pollution and other forms of environmental degradation.

In the Niger Delta, externalisation is an extension of dispossession as polluted water sources, fields and fisheries are simply lost to their owners. But the effects are not restricted to this. The health impacts of air pollution spread across a wide area, and all who rely on locally produced food – whether from their own production or bought at market – risk contamination. On a global scale, the emissions of carbon dioxide (CO_2) and methane from Nigeria's flares make a substantial contribution to climate change and the costs will, again, fall heaviest on the poor.

Nigeria does have environmental laws that should notionally ensure that these costs are internalised – that they are actually paid by the corporations. The state does not, however, have the capacity or the inclination to enforce the law. This contrasts starkly with the political will and resources devoted to enforcing dispossession. Consequently, corporations have been almost entirely self-regulating in respect of their environmental practices in Nigeria and have externalised costs without inhibition. This began to change in the 1990s when the actions of local people's movements combined with international civil society organisations to expose corporate practices. The corporations, notably Shell, perceived this primarily as a public relations disaster and responded mostly with public relations 'spin'. Whatever caution they now exercise is proportional to the national, and particularly the international, visibility of their practice.

Chad is Africa's newest oil state and the externalised costs to date have been mainly those associated with exploration, drilling and construction. It has also been subject to unusual scrutiny as the political price that the World Bank paid for insisting that here it would demonstrate how oil extraction can contribute to alleviating poverty – even against the odds – was extremely high. The oil started to flow in 2003 and the flares are burning above the villages of southern Chad. The impacts will be felt in time and will most likely escalate. They will be mitigated only in so far as it is possible for local and international civil society to maintain present levels of scrutiny.

Exclusion

Exclusion relates to decision-making power in the market and in society. Given the weight of economic forces in shaping broader social institutions and relations, these two aspects of exclusion frequently reinforce each other. The institutions of the market are specifically designed to remove decision-making from the public sphere and so exclude all who do not have an interest in profit. Thus, those who are dispossessed, or who carry the externalised costs of production, are prevented from contesting the theft or contamination of their resources.

Niger Delta communities have a long history of resisting the enclosure of their land. The Movement for the Survival of the Ogoni People (MOSOP) became the best-known organisation of resistance and an inspiration for communities across the Delta. In 1993, it organised mass protests throughout Ogoniland and forced Shell to close down its Ogoni production wells, although active pipelines still cross the territory.

Resistance was met with brutal repression. It started with security force attacks thinly disguised as inter-ethnic violence. At the same time, Shell was trying to buy off MOSOP leaders. Then, in 1994, four 'moderate' Ogoni chiefs were murdered at Giokoo. The circumstances indicate that they were killed by security operatives acting under cover. Prominent MOSOP leaders were immediately accused of the murders and arrested – without allowing time even for the pretence of an investigation. In 1995, Ken Saro-Wiwa and eight others were executed on the order of a rigged court.

The use of brutal security force violence did not begin or end in Ogoniland. From the early 1990s protests across the Delta became more organised and numerous ethnic groups adopted charters loosely modelled on the Ogoni Bill of Rights. They commonly claimed the right to control land and natural resources, including oil, and demanded a meaningful political voice within a restructured Nigerian federation.

The savagery of the security force response also intensified throughout the decade. Ijaw youth greeted the new year of 1999 by mobilising in support of the Ijaw Youth Council's Kaiama Declaration. In response, security forces killed over one hundred people and burned down between ten and twenty homes. In many similar incidents around the Delta, corporate helicopters and boats were seen carrying security forces. The corporations routinely deny involvement. However, new evidence brought to light in preparation for a court case against Chevron indicates that soldiers not only used Chevron's helicopters in a 1999 attack on the villages of Opia and Ikenyan, but that Chevron paid them for the operation.

The death of military dictator Sani Abacha in 1998 opened the way to a restoration of civilian rule and the election of Olusegun Obasanjo, himself a former military ruler, as president. The occupation of Ogoniland was lifted, but the Delta is still saturated with security forces and abuse of people is routine. On the other side, people have occupied oil facilities and forced temporary shut-downs across the Delta. In 2002, several hundred women occupied Chevron-Texaco's Escravos terminal in the Delta state for ten days, one example of the growing assertiveness of women in resistance.

In this period, gun-trafficking in the Delta has escalated and armed youth groups, sometimes known as 'cults' or 'area boys', have emerged. Mostly, it appears that they have been armed by politicians to intimidate opposition party supporters, by local elites to secure their control over oil subcontracts and payoffs against rival factions, or through 'illegal bunkering' networks responsible for the wholesale theft of oil. Cult leaders have also been used to infiltrate and subvert resistance movements.

Chad's president, Idriss Deby, took power when his rebel troops captured the capital N'Djamena in 1990, but subsequently gave his regime a veneer of democratic legitimacy through rigged elections. Friends of the Earth report that, in 1997 and 1998, hundreds of civilians were massacred in the project area by national troops, for the sake of 'pacifying' the region to make way for oil development. Community consultations were conducted in the presence of security forces at least until 1997 and thereafter in the presence of government officials.

Chad ranks at the bottom of international league tables on most indicators of good governance, including corruption and voice and accountability. Government officials are increasingly appointed from a narrow clique around the president and arbitrary arrests, torture and summary executions by security forces are routine. Independent radio stations are regularly closed down and journalists arrested in response to critical broadcasts. In 2003, a station run by local human rights groups was closed down less than two weeks after international VIPs had been in the country for the October 2003 pipeline inauguration.

These then are the means by which the value subtraction chain is made to work. A detailed documentation of every incident of abuse would be a very long book indeed. Our intention here is simply to show the oil industry at work and the stories told are only a very small selection of the stories that could be told.

8

Oil Companies and African Wealth Depletion

Patrick Bond

Who will buy carbon credits, which permit ongoing carbon dioxide (CO_2) emissions? Aside from governments – of which the Netherlands is the most important, supporting 16 per cent of the carbon market – the primary beneficiary of emissions trading will be Big Oil, namely the huge corporations which are the main contributors to global warming. The oil majors have already invested a great deal of money and promotional work in establishing a non-transparent, profoundly flawed system that will permit Kyoto Protocol emissions reduction targets to be foiled with fake carbon offsets.

These same companies have recorded windfall profits as the oil price soared from around US$10 per barrel in 1998 to more than US$150 per barrel at peak in 2008. What kinds of policies are they now investing in, and not merely with respect to a dangerous new carbon market? What other forms of extra-economic support can they claim?

Big oil companies are the beneficiaries of a formidable rise in militarism across the world, including Africa. With genocidal and dictatorial regimes such as Sudan and Equatorial Guinea hosting Big Oil, politicians in Washington, DC, Beijing, Brussels and Tshwane (formerly Pretoria) suddenly forget their human rights and democracy rhetoric.

Likewise, we must ask, are the oil and other extractive firms that are responsible for so much greenhouse gas emission paying the full price for looting Africa of its resources? As shown below, even the World Bank now concedes that the depletion of non-renewable resources by large corporations has not been matched by new investments or revenues. The citizenry of many African countries are much worse off in quantitative terms, after the plague of locusts we know as the oil majors. In qualitative respects, not dealt with here, the corruption and related problems associated with oil extraction are also a severe resource curse.

For these reasons and because Northern production and consumption habits are creating disastrous climate change, the case for ecological debt – owed by the North (and South Africa) to the South – should be very forcefully raised.

Imperialism, oil and Africa

The period following Washington, DC's failed early 1990s Somali intervention, when the Pentagon's warriors let Africa slip from view, may have come to an end on 11 September 2001. Between the search for oil and terrorist threats – bearing in mind that the African National Congress (ANC) was listed by the US state department as 'terrorist' until the 1990s – the US military-petroleum-industrial complex has increasing reason to occupy Africa.

US Army General Charles Wald, who controls the Africa Programme of the European Command, told the British Broadcasting Corporation (BBC) in early 2004 that he aims to have five brigades with 15 000 men working in co-operation with regional partners, including South Africa, Kenya, Nigeria and two others still to be chosen (Plaut 2004).

The North Atlantic Treaty Organisation's (NATO's) Supreme Allied Commander for Europe, General James Jones, confirmed the US geographical strategy in May 2003: 'The carrier battle groups of the future and the expeditionary strike groups of the future may not spend six months in the Mediterranean Sea but I'll bet they'll spend half the time down the West Coast of Africa'.[1] Within weeks, 3 000 US troops had been deployed off the coast of Liberia, and went briefly ashore to stabilise the country after Charles Taylor departed.

Potential US bases were suggested for Ghana, Senegal and Mali, as well as the North African countries of Algeria, Morocco and Tunisia (*Ghana News*, 11 June 2003). Another base was occupied by 1 500 US troops in the small country of Djibouti. Botswana and Mozambique were also part of the Pentagon's strategy and South Africa would remain a crucial partner. In September 2006, Pentagon head Donald Rumsfeld announced that a special 'Africa command' would be established.

Central and Eastern Africa remain problem areas and not merely because of traditional French and Belgian neocolonial competition with British and US interests (Taylor 2003: 49). President Clinton's refusal to cite Rwanda's situation as formal genocide in 1994 was an infamous failure of nerve in terms of the emerging doctrine of humanitarian imperialism – and in stark contrast to intervention in the (white-populated) Balkans.

With an estimated three million people dead in Central African wars, partly due to struggles over access to coltan and other mineral riches, conflicts worsened between and within the Uganda/Rwanda bloc, vis-à-vis the revised alliance of Laurent Kabila's Democratic Republic of the Congo (DRC), Zimbabwe, Angola and Namibia. Only with Kabila's assassination in 2001 and the South African government's management of uneasy elite deals in the DRC and Burundi, did matters settle, however briefly, into a fragile peace, combining neoliberalism with opportunities for minerals extraction.

However, as turmoil resumed in mid-2004, it was clear that coups and outbreaks of strife would be a constant threat, demonstrating how precarious the South African government's deals are when deeper tensions remain unresolved. Another particularly difficult site is Sudan, where US Delta Force troops have been sighted in informal operations, perhaps because although China showed some interest in oil exploration there during the country's civil war chaos, US oil firms have subsequently arrived. Both US and South African oil deals with Sudan may explain why their interventions against the Darfur genocide are so half-hearted.

On the west coast, the major petro-prize remains the Gulf of Guinea and again Equatorial Guinea's deals with South Africa stand out as particularly obnoxious, given the regime's dictatorial practices. South Africa's oil deals with Iraq and Nigeria have been exposed as fraught with ethical problems, but it is the United States that has the power and interests to make a much greater mess. With oil shipment from Africa to Louisiana refineries taking many fewer weeks than from the Persian Gulf, the world's shortage of super-tankers is eased by direct sourcing from West Africa's offshore oil fields.

In this context, it is not surprising that of US$700 million destined to develop a 75 000-strong United Nations (UN) peacekeeping force in coming years, US$480 million is dedicated to African soldiers, nor that South Africa also recently bought into US-aligned military strategies for troop deployment.[2] But Africa is also a site for the recruitment of well-tested private mercenaries. An estimated 1 500 South Africans – including half of Mbeki's own 100 personal security force – joined firms such as South Africa's Executive Outcomes and British-based Erinys to provide more than 10 per cent of the bodyguard services in occupied Iraq (*Vancouver Sun*, 11 May 2004). Some African countries, including Eritrea, Ethiopia and Rwanda, joined the Coalition of the Willing against Iraq in 2003, although temporary UN Security Council members Cameroon, Guinea and the Republic of the Congo opposed the war, in spite of bullying from the US government.

Also on the US geopolitical front, the Central African Republic proved reliable during the reconciliation of Jacques Chirac and the Bush regime in March 2004, when Haitian president Jean-Bertrand Aristide was kidnapped and temporarily dumped there, prior to taking up a cautious residence in South Africa. Africa is also an important site for Washington, DC's campaigns against militant Islamic networks, especially in Algeria and Nigeria in the northwest, Tanzania and Kenya in the east and South Africa.

Control of African immigration to the United States and Europe is crucial, in part through the expansion of US-style incarceration via private-sector firms such as Wackenhut, which has invested in South African privatised prison management, along with the notorious Lindela extradition camp for illegal immigrants, part of a highly racialised global detention and identification system.

Of course, the US military machine does not roll over Africa entirely unimpeded. Minor roadblocks have included South Africa's rhetorical opposition to the Iraq war, conflicts within the UN Human Rights Commission (especially over Zimbabwe) and the controversy over US citizens' extradition to the International Criminal Court (ICC). On the eve of Bush's 2003 Africa trip, the Pentagon announced that it would withdraw US$7.6 million worth of military support to South Africa, because the government – along with 34 military allies of Washington, DC (and 90 countries in total) – had not agreed to give US citizens immunity from prosecution at The Hague's new International Criminal Court. Botswana, Uganda, Senegal and Nigeria, also on Bush's itinerary, signed these blackmail-based immunity deals and retained US aid (SAPA, 2 July 2003).[3]

After Bush returned home, in mid-July, the US House of Representatives extended a ban on military assistance to 32 countries, including South Africa. But Washington, DC's ambassador to Tshwane, Cameron Hume, quickly announced that several bilateral military deals would go ahead in any case. According to Peter McIntosh of *African Armed Forces* journal, the United States 'had simply re-routed military funding for South Africa through its European Command in Stuttgart'. Hume reported the Pentagon's desire 'to train and equip two additional battalions to expand the number of forces the [South African National Defence Force] have available for peacekeeping in Africa'. South African newspaper *This Day* commented, in the wake of two successful joint US/South African military manoeuvres in 2003–04: 'Operations such as Medflag and Flintlock clearly have applications other than humanitarian aid, and as the US interventions in Somalia and Liberia have shown, humanitarian aid often requires forceful protection' (Schmidt 2004).

The two countries' military relations were fully 'normalised' by July 2004, in the words of South African deputy-minister Aziz Pahad. In partnership with General Dynamics Land Systems, state-owned Denel immediately began marketing 105-millimetre artillery, alongside a turret and light armoured vehicle hull, in support of innovative Stryker Brigade Combat Teams ('a 3 500-personnel formation that puts infantry, armour and artillery in different versions of the same 8×8 light armoured vehicle'). According to one report, 'The turret and gun is entirely proprietary to Denel, using only South African technology. At sea level, it can fire projectiles as far as 36 km' (SAPA 2004a). This followed a period of serious problems for the South African arms firm and others like it (Armscor and Fuchs), which were also allowed full access to the US market in July 2004 after paying fines for apartheid-era sanctions-busting (Batchelor and Willett 1998: 329–42).

Given the South African government's 1998 decision to invest US$6 billion in mainly offensive weaponry such as fighter jets and submarines, there are growing fears that peacekeeping is a cover for a more expansive geopolitical agenda and that Mbeki is tacitly permitting a far stronger US role in Africa – from the oil-rich Gulf of Guinea and Horn of Africa, to training bases in the south and north – than is necessary (Black 2004). On the surface, the South African government's senior roles in the mediation of conflicts in Burundi and the DRC in 2003 appeared positive. However, closer to the ground, the agreements more closely resemble the style of elite deals that lock in place 'low-intensity democracy' and neoliberal economic regimes. Moreover, because some of the belligerent forces were explicitly left out, the subsequent weeks and months after declarations of peace witnessed periodic massacres of civilians in both countries and a near coup in the DRC.

In the light of the effective geopolitical and military alliance between the South African and US governments, how are we to interpret the South African government's recent global political zigzag? On the one hand, Tshwane's grand continental plan, the New Partnership for Africa's Development (NEPAD), was declared 'philosophically spot-on' by the Bush regime (Gopinath 2003) and Mbeki was anointed Washington, DC's 'point man' for resolving the Zimbabwe crisis by Bush himself, during the US president's July 2003 visit to South Africa, in spite of Mbeki's continual nurturing of Mugabe's repression. Hence in January 2003, Nelson Mandela remarked, 'If there is a country which has committed unspeakable atrocities, it is the United States of America' (CBS News 2003), but in May 2004, he retracted his criticism, simply because 'The United States is the most powerful state in the world, and it is not good to remain in tension with the most powerful state' (CNN 2004). As Greg Mills of the South African Institute of International Affairs explains:

> I think there was a bluster by the South African government, or those associated near or around it, prior to the American invasion of Iraq in March last year, but that was toned down fairly quickly by the South African government and most notably, President Mbeki. Really, there has not been much in the way of condemnation of the American position since March last year. (Williams 2004)

Competition from other neocolonial sponsors has occasionally been a factor limiting US arrogance, for example in the only partially successful attempt by Monsanto to introduce genetically modified agriculture in Africa. Zambia, Zimbabwe and Angola have rejected World Food Programme and US food relief because of fears of future threats to their citizens and, not coincidentally, to European markets. Linking its relatively centralised aid regime to trade through bilateral regionalism, the European Union (EU) aims to win major Africa-

Caribbean-Pacific (ACP) country concessions on investment, competition, trade facilitation, government procurement, data protection and services, which along with grievances over agriculture, industry and intellectual property were the basis of ACP withdrawal from Cancun.

The European Union's economic partnership agreements (EPAs), under the Cotonou Agreement (which replaced the Lome Convention), will signify a new, even harsher regime of reciprocal liberalisation to replace the preferential agreements that tied so many African countries to their former colonial masters via cash-crop exports. If the EPAs are agreed upon, what meagre organic African industry and services that remained after two decades of structural adjustment will probably be lost to European scale economies and technological sophistication.

An April 2004 meeting of parliamentarians from East Africa expressed concern, 'that the pace of the negotiations has caught our countries without adequate considerations of the options open to us, or understanding of their implications, and that we are becoming hostage to the target dates that have been hastily set without the participation of our respective parliaments'. Even Botswana's neoliberal president Festus Mogae admitted, 'We are somewhat apprehensive towards EPAs despite the EU assurances. We fear that our economies will not be able to withstand the pressures associated with liberalisation.'[4] But the European Union's substantial aid carrots and sticks will be the final determinant, overriding democratic processes. Fortunately, South Africa opposed EPAs by 2008, thanks to strong leadership by deputy trade minister Rob Davies.

What of Washington, DC's development aid to Africa? During the early 1990s, numerous US Agency for International Development mission offices in Africa were closed by the Clinton administration. The highest-profile measures now relate to HIV/AIDS treatment, amounting to what the state department called its 'full-court press' – including threats of further aid cuts – against governments that made provisions for generic medicines production, which Clinton only backed away from in late 1999 because of sustained activist protest (Bond 1999: 4). Bush promised a US$15 billion AIDS programme, then whittled it down to a fraction of that, then refused to provide funds to the UN Global Fund to Fight AIDS, TB and Malaria, and then prohibited US government financing of generic medicines. Bush also introduced an innovative vehicle to fuse neoliberal market conditionality with, supposedly, greater social investment: the Millennium Challenge Account (MCA).

With US aid budgets still declining in real terms, the delinked MCA funding will rise from US$1 billion in 2004 to US$5 billion in 2006, a 100 per cent increase on 2004 spending for all US overseas development assistance. But of 74 low-income countries that are meant to be eligible, of which 39 are from Africa, only 16 passed the first test of governance and economic freedom in May 2004. Half of these were African: Benin, Cape Verde, Ghana, Lesotho,

Madagascar, Mali, Mozambique and Senegal. The criteria for funding these countries' aid programmes have been established by a series of think tanks and quasi-government agencies: Freedom House (civil liberties and political rights), the World Bank Institute (accountability, governance and control of corruption), the International Monetary Fund (IMF) and the Heritage Foundation Index of Economic Freedom (credit ratings, inflation rates, business start-up times, trade policies and regulatory regimes) and the World Health Organisation and UN (public expenditure on health and primary education, immunisation rates and primary school completion rates) (cited in *SA Institute for International Affairs e-Africa*, May 2004).[5] Washington, DC's attempt to disguise and legitimise imperialism through aid that carries good governance and social investment conditionalities dates back to the Clinton era, but under Bush's MCA it involves more sophisticated disciplinary neoliberal surveillance, especially in combination with the World Bank (Alexander 2004).

However, with so few African states receiving MCA funding and so much more at stake than can be handled by the expansion of military spending, it is vital for Washington, DC to identify reliable allies in Africa to foster both imperialist geopolitics and neoliberal economics. Does South Africa qualify? South Africa is, in many ways, playing a sub-imperial role in Africa. This has become acutely obvious in terms of energy, not only through Eskom's reach up the continent, but also through new oil agreements with two of the world's most vicious governments: Sudan and Equatorial Guinea. A 2005 study by Human Sciences Research Council (HSRC) officials John Daniel and Jessica Lutchman bears this out:

> South Africa is dependent for approximately 98 per cent of its crude oil needs on imports. Of this, 75 per cent is imported from the Middle East and 23 per cent from Africa. This latter figure reflects a considerable increase in recent years. In 2001, for example, African imports stood at only nine per cent. Its main African supplier is Nigeria with the others being Angola, Cameroon and Gabon. In the last five years, however, South Africa – or more precisely South African companies operating with state support – has moved to reduce its import dependency through a process of buying into the African oil market as either a sole proprietor or in a partnership arrangement. In pursuit of this goal, South Africa has employed a combination of economic muscle, technical edge and tactical diplomacy (e.g President Mbeki's agreement with Sudan and former Deputy President Zuma's state visit to Angola). Leading the attempt to secure this greater stake in African crude oil production have been Sasol and the CEF-owned PetroSA. PetroSA has agreements with Algeria, Angola, Gabon, Nigeria and Sudan. Sasol on the other hand is operating in Gabon and Equatorial Guinea . . .

> Access to Africa's range of energy resources will be a key to the success of South Africa's developing energy strategy. This sector is already contested terrain and other energy-deficit states are eyeing the African market. In its scramble to acquire a share of this market, the ANC government has abandoned any regard to those ethical and human rights principles which it once proclaimed would form the basis of its foreign policy. Its approach in regard to energy is one in which the national interest is being interpreted purely on material grounds. (Daniel and Lutchman 2005)

Unpacking investments in African oil fields

Who is South Africa competing with and how? Given the appearance of growing foreign direct investment in Africa, a great deal of nuance is required in deconstructing the data, especially from 1997, for it appears that the peaks are associated with special circumstances. The Angolan 1999 oil investment peak was limited to the offshore Cabinda fields, while on the Angolan mainland, a repressive, corrupt state regime waged war against a right-wing guerrilla army. The 1990s investments in Nigerian oil occurred largely under Sani Abacha's 1990s dictatorial rule and were offset by his looting of state resources to private Swiss and London accounts. And the other peak of foreign investment, into South Africa, reflects both the failed telecommunications privatisation of 1997 and statistical accounting changes associated with the relisting of the country's largest firms to London.

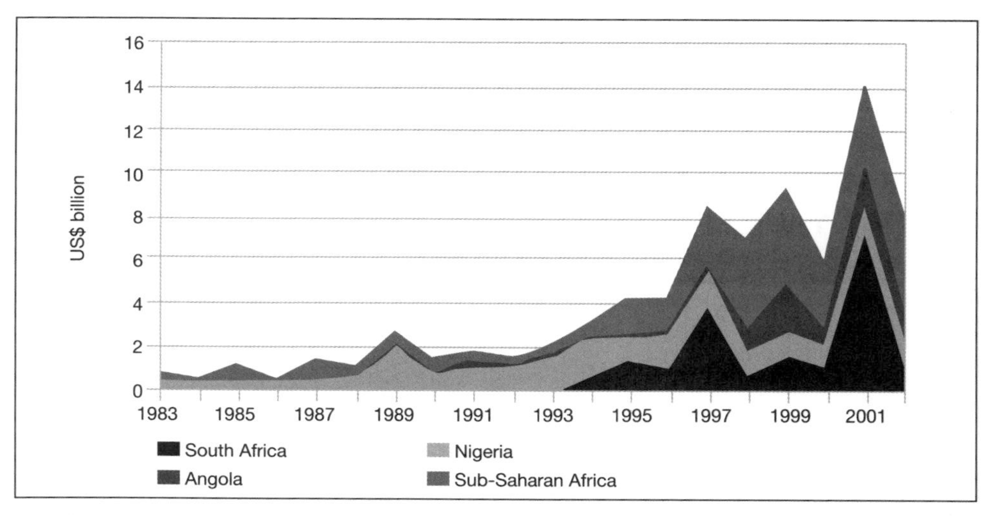

Rising foreign investment in Africa.

Source: The Blair Africa Commission

The oil sector is a clear case in which profit and dividend outflows, often lubricated by corruption, have had extremely negative consequences. As demonstrated by the Open Society-backed campaign, 'Publish what you Pay', elites in Africa's oil-producing countries – Angola, Chad, Congo, Equatorial Guinea, Gabon, Nigeria and Sudan – are among the world's least transparent.[6] In Nigeria, demands by the Ogoni people relate not only to the massive destruction of their Delta habitat, but also to the looting of their natural wealth by Big Oil.[7] In all these respects, diverse forces in society have moved from considering oil merely a matter of private property, to be negotiated between corporations and governments, as was the case during much of the twentieth century. Instead, these forces now treat oil as part of a general commons of a national society's natural resources. George Caffentzis explains:

> There are three levels of claims to petroleum as common property, correlating with three kinds of allied communities that are now taking shape, for there is no common property without a community that regulates its use:
>
> - first, some local communities most directly affected by the extraction of petroleum claim to own and regulate the petroleum under its territory as a commons;
> - second, Islamic economists claim for the Islamic community of believers, from Morocco to Indonesia, and its representative, the 21st century Caliphate in formation, ownership of and the right to regulate the huge petroleum fields beneath their vast territory;
> - third, UN officials claim for the 'coming global community' the right to regulate the so-called global commons-air, water, land, minerals (including petroleum) and 'nous' (knowledge and information). This imagined global community is to be represented by a dizzying array of 'angels' that make up the UN system, from NGO activists to UN environmentalist bureaucrats to World Bank 'green' advisors. (Caffentzis 2004)

If we take as given that there is some merit in considering natural resources as a commons, their depletion plus associated negative externalities – such as the social devastation caused by mining operations – must be taken seriously. In the case of South Africa, for example, the value of natural minerals capital in the soil fell from US$112 billion in 1960 to US$55 billion in 2000, according to a United Nations Development Programme (UNDP) estimate (2004).

The World Bank has even addressed the issue of natural resources in a 2005 document, 'Where is the Wealth of Nations? Measuring Capital for the 21st Century'. The Bank methodology for correcting bias in gross domestic product (GDP) wealth accounting is nowhere near as expansive as that, for instance, of the San Francisco group Redefining Progress, which estimates that global GDP has been declining in absolute terms since the mid-1970s, if natural resource depletion, pollution and a variety of other factors are accounted for.[8]

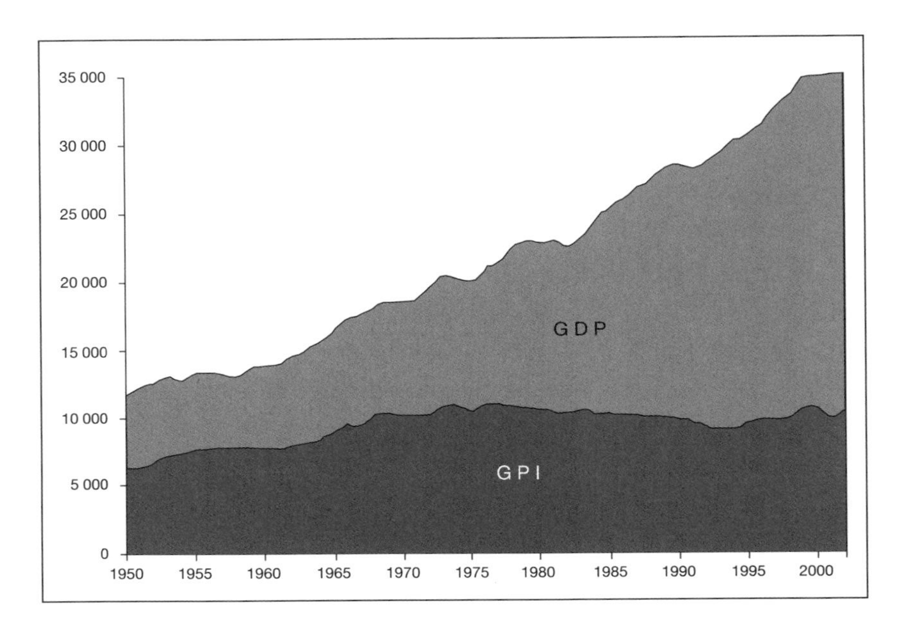

Global GDP versus a genuine progress indicator, 1950–2003.

Source: http://www.redefiningprogress.org

Direct investment (DI) may not contribute to net GDP growth if resource depletion and pollution associated with extractive industries are factored in. In the case of Bolivia, for example, the Bank's first-cut method subtracts from an existing 12 per cent savings/gross national income (GNI) the following: fixed capital depreciation, depletion of natural resources and pollution (while increasing savings based on education expenditure), leaving a net –3.5 per cent savings/GNI rate.

It should be noted here that in making estimates about the decline in a country's wealth due to energy, mineral or forest-related depletion, the World Bank has a minimalist definition based on international pricing (not potential future values when scarcity becomes a more crucial factor, especially in the oil industry). The Bank does not calculate the damage done to the local environment, to workers' health or safety and especially to women in

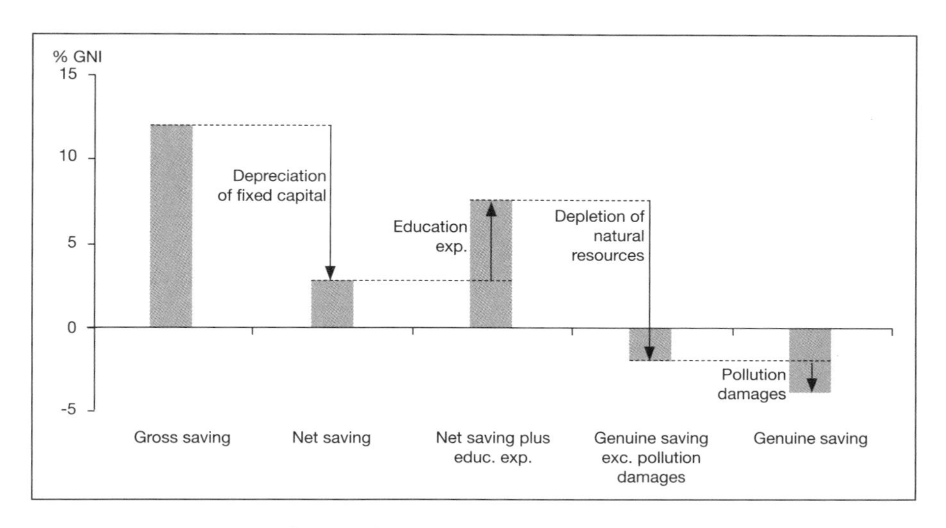

World Bank methodology for 'genuine saving' calculations.

Source: World Bank 2005

communities around mines. Moreover, the Bank's use of average – not marginal – cost resource rents also probably leads to underestimations of the depletion costs.

However, the methodology at least indicates some of the trends associated with foreign direct investment (FDI) related resource extraction. In particular, the attempt to generate a 'genuine savings' calculation requires adjusting net national savings to account for resource depletion. The Bank suggests the following steps:

> From gross national saving the consumption of fixed capital is subtracted to give the traditional indicator of saving; net national savings. The value of damages from pollutants is subtracted. The pollutants carbon dioxide and particulate matter are included. The value of natural resource depletion is subtracted. Energy, metals and mineral and net forest depletion are included. Current operating expenditures on education are added to net national saving to adjust for investments in human capital. (World Bank 2005)

Naturally, given oil extraction, the Middle East and North Africa have the world's most serious problem of net negative GNI and savings under this methodology. But Africa is second worst, with several years during the early 1990s showing net negative GNI, once extraction of natural resources is factored in.

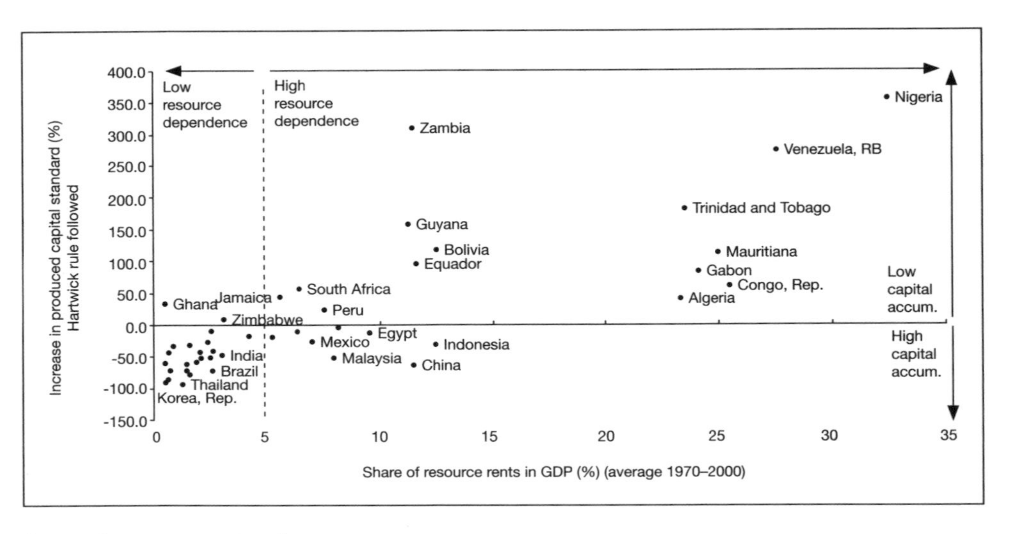

Dependence upon extractive resources.

Source: World Bank

Hence for every percentage point increase in a country's extractive-resource dependency, that country's potential GDP falls by 9 per cent (as against the real GDP recorded) according to the Bank. The African countries most affected – i.e., with high resource dependence and low capital accumulation – include Nigeria, Zambia, Mauritania, Gabon, Congo, Algeria and South Africa.

An even more nuanced breakdown of a country's estimated tangible wealth is required to capture not only obvious oil-related depletion and rent outflows, but also other subsoil assets, timber resources, non-timber forest resources, protected areas, cropland and pastureland. The produced capital normally captured in GDP accounting is added to the tangible wealth. In the case of Ghana, this amounted to US$2 022 per person in 2000. The same year, the gross national saving (GNS) of Ghana was US$40 and education spending was US$7. These figures require downward adjustment to account for the consumption of fixed capital (US$19) and the depletion of energy (US$0), mineral (US$4) and net forest wealth (US$8). The adjusted net saving was, hence US$16 per person and given a population growth of 1.7 per cent, that brought per capita wealth down by US$18 per person in 2000.

In the case of Ghana, US$12 of the US$18 decline can be attributed to minerals and forest related depletions and it is unclear how much of that is a product of transnational capital extracting these resources from Ghana. The largest indigenous (and black-owned) mining firm in Africa, Ashanti, was subsequently taken over by AngloGold, so it is safe to

African countries' adjusted national wealth and 'savings gaps', 2000.

Country	Income per capita (US$)	Population growth rate (%)	Adjusted net saving per capita (US$)	Change in wealth per capita (US$)
Benin	360	2.6	14	–42
Botswana	2 925	1.7	1 021	814
Burkina Faso	230	2.5	15	–36
Burundi	97	1.9	–10	–37
Cameroon	548	2.2	–8	–152
Cape Verde	1 195	2.7	43	–81
Chad	174	3.1	–8	–74
Comoros	367	2.5	–17	–73
Rep. of Congo	660	3.2	–227	–727
Côte d'Ivoire	625	2.3	–5	–100
Ethiopia	101	2.4	–4	–27
Gabon	3 370	2.3	–1 183	–2 241
The Gambia	305	3.4	–5	–45
Ghana	255	1.7	16	–18
Kenya	343	2.3	40	–11
Madagascar	245	3.1	9	–56
Malawi	162	2.1	–2	–29
Mali	221	2.4	20	–47
Mauritiania	382	2.9	–30	–147
Mauritius	3 697	1.1	645	514
Mozambique	195	2.2	15	–20
Namibia	1 820	3.2	392	140
Niger	166	3.3	–10	–83
Nigeria	297	2.4	–97	–210
Rwanda	233	2.9	14	–60
Senegal	449	2.6	31	–27
Seychelles	7 089	0.9	1 162	904
South Africa	2 837	2.5	246	–2
Swaziland	1 375	2.5	129	8
Togo	285	4.0	–20	–88
Zambia	312	2.0	–13	–63
Zimbabwe	550	2.0	53	–4

Source: World Bank 2005: 66

assume than an increasing amount of Ghana's wealth flows out of the country, leaving a net negative per capita tangible wealth for Ghanaians.

Other mining houses active in Africa that had their roots here – Lonrho, Anglo, DeBeers, Gencor/Billiton – are also now based offshore. While this makes calculating the outflow from Africa relatively easier, the drive by London, New York and Sydney shareholders for profits means accumulation of capital within Africa is stymied.

Other African countries whose economies are primary product dependent fare much worse, according to the Bank methodology. In the worst case, Gabon's people lost US$2 241 each in 2000, as oil companies depleted the country's tangible wealth. The Republic of the Congo (–US$727), Nigeria (–US$210), Cameroon (–US$152), Mauritania (–US$147) and Côte d'Ivoire (–US$100) are the other African countries whose people lost more than US$100 in tangible national wealth each in 2000 alone. A few countries did benefit, according to the Bank's tangible wealth measure, including the Seychelles (+US$904), Botswana (+US$814) and Namibia (+US$140), but the vast majority of African countries saw their wealth depleted.

Even Africa's largest economy, South Africa, which from the early 1980s has been far less reliant upon minerals extraction, recorded a US$2 drop in per capita wealth in 2000 using this methodology. According to the World Bank, the natural wealth of US$3 400 per person in South Africa included subsoil assets (worth US$1 118 per person); timber (US$310); non-timber forest resources (US$46); protected areas (US$51); cropland (US$1 238) and pastureland (US$637). This sum can be compared to the value of produced capital (plant and equipment) and urban land (together worth US$7 270 per person in 2000). Hence even in Africa's most industrialised economy, the estimated value of natural resources is nearly half of the measurable value of plant, equipment and urban land. Given the constant depletion of its natural resources, South Africa's official GNS rate of 15.7 per cent therefore should be adjusted downwards. By subtracting consumption of fixed capital at 13.3 per cent, the net national savings is actually 2.4 per cent, added to which should be education expenditure (among the world's highest) at 7.5 per cent. Then subtract mineral depletion of 1 per cent; forest depletion of 0.3 per cent; pollution ('particulate matter') damage of 0.2 per cent and (probably undervalued) CO_2 damage of 1.6 per cent. In total, the actual 'genuine savings' of South Africa is reduced to just 6.9 per cent of national income.

How much of this deficit from the 15.7 per cent savings rate can be attributed to foreign investors? Not only is mineral depletion biased to benefit overseas mining houses, but the CO_2 damage is largely done by the smelters owned by large multinational corporations, including Mittal Steel, BHP Billiton (Alusaf) and the Anglo group. The amount is substantive and further estimates should reasonably be made.

Carbon emissions and ecological debt

CO_2 emissions damage is, essentially, a draw-down from the global commons. During the early 1990s, the idea of the North's ecological debt to the South began gaining currency in Latin America, thanks to non-governmental organisations (NGOs), environmentalists and politicians (including Fidel Castro of Cuba and Virgilio Barco of Colombia). According to Spanish ecological economist Joan Martinez-Alier:

> The notion of an ecological debt is not particularly radical. Think of the environmental liabilities incurred by firms (under the United States Superfund

> legislation), or of the engineering field called 'restoration ecology', or the proposals by the Swedish government in the early 1990s to calculate the country's environmental debt. Ecologically unequal exchange is one of the reasons for the claim of the Ecological Debt. The second reason for this claim is the disproportionate use of Environmental Space by the rich countries. (Martinez-Alier 2003)[9]

In the first category, Martinez-Alier lists:

- unpaid costs of reproduction or maintenance or sustainable management of the renewable resources that have been exported;
- actualised costs of the future lack of availability of destroyed natural resources;
- compensation for, or the costs of reparation (unpaid) of the local damages produced by exports (for example, the sulphur dioxide of copper smelters, the mine tailings, the harm to health from flower exports, the pollution of water by mining), or the actualised value of irreversible damage;
- (unpaid) amount corresponding to the commercial use of information and knowledge on genetic resources, when they have been appropriated gratis ('biopiracy'). For agricultural genetic resources, the basis for such a claim already exists under the United Nations Food and Agriculture Organisation's (FAO's) Farmers' Rights.

In the second, he cites 'lack of payment for environmental services or for the disproportionate use of Environmental Space':

- (unpaid) reparation costs or compensation for the impacts caused by imports of solid or liquid toxic waste; and
- (unpaid) costs of free disposal of gas residues (carbon dioxide, CFCs, etc), assuming equal rights to sinks and reservoirs.

These aspects of ecological debt defy easy measurement. Each part of the ecological balance sheet is highly contested and information is imperfect. As Martinez-Alier shows in other work, tropical rainforests used for wood exports have an extraordinary past that we will never know and ongoing biodiversity whose destruction we cannot begin to value. However, he acknowledges, 'although it is not possible to make an exact accounting, it is necessary to establish the principal categories [of ecological debt] and certain orders of magnitude in order to stimulate discussion' (Martinez-Alier 1998). The sums involved are potentially vast. Just to take the case of CO_2 emissions, according to Martinez-Alier,

> Jyoti Parikh (a member of the UN International Panel on Climate Change) [argues that] if we take the present human-made emissions of carbon, the average is about one ton per person per year. Industrialised countries produce three-fourths of these emissions, instead of the one-fourth that would correspond to them on the basis of population. The difference is 50 per cent of total emissions, some 3000 million tons. Here the increasing marginal cost of reduction is contemplated: the first 1000 million tons could be reduced at a cost of, say, $15 per ton, but then the cost increases very much. Let us take an average of $25: then a total annual subsidy of $75 billion is forthcoming from South to North. (Parikh 1995, cited in Martinez-Alier 1998)

Excess use of the planet's CO_2 absorption capacity is merely one of the many ways that the South is being exploited by the North on the ecological front. Africans are most exploited in this regard because non-industrialised economies have not begun to utilise more than a small fraction of what should be due under any fair framework of global resource allocation.

Conclusion

As we have seen, the overlapping and interlocking roles of imperialism, South African sub-imperialism and extractive FDI in oil- and resource-rich countries are closely related to the North's – and South Africa's – fossil-fuel addiction. Hence any genuine accounting of national economic welfare stemming from further extractive investments should take into account not only the danger of imperial intervention and the financing of repressive dictatorships, but also the net negative impact on national wealth, including natural resources. The new World Bank accounting of genuine savings that partially takes into account depletion of natural resources by foreign corporations is a helpful innovation. To take the methodology forward and rigorously estimate an Africa-wide extraction measure, to account for the way extractive FDI generates net negative welfare or savings, still remains as an exercise.

But the overall conclusion is sufficiently disturbing that a very different conception of oil wealth and politics begins to emerge. Indeed, given what powerful forces from Washington, DC to Brussels to Tshwane and in between have shown themselves capable of, it may well make sense for African oil to stay in the soil until bottom-up democratisation allows for a more serious cost-benefit analysis to be done. Meanwhile, if civil society in Africa is to continue fighting against extractive industries on behalf of human, women's, workers and environmental rights – as we are taught so often from the Nigerian Delta to Chad to Botswana – then the rest of the world's conscientised people must increase their own efforts to repay the ecological debt owed to Africa. Such a turn, uniting the world's people against climate change and myriad forms of local destruction caused by Big Oil and financial institutions such as the World Bank, increasingly enforced by the Pentagon and the South African National Defence Force, would be a welcome direction for internationalism.

It is not an entirely pessimistic story, though. While the 24 September 2005 protests of 300 000 US residents against the Iraq War and IMF/World Bank annual meetings in Washington, DC proved that what the *New York Times* termed the world's 'second superpower' – informed public opinion – is still actively hostile to the White House, South African civil society's Anti-War Coalition, Jubilee movement, anti-privatisation social movements, progressive Congress of South African Trade Unions (COSATU) trade unions and other forces also have further obligations to connect the dots and hit the streets.

Notes

1. See http://www.allAfrica.com.
2. The major dilemma, here, appears to be the very high level of HIV-positive members of the armed forces in key countries. See Elbe 2003: 23–44.
3. Other African countries where US war criminals are safe from ICC prosecutions thanks to military-aid blackmail are the DRC, Gabon, The Gambia, Ghana, Kenya, Mauritius, Sierra Leone and Zambia.
4. See http://www.epawatch.net/general/text.php?itemID=161&menuID=28 and http://www.twnafrica.org/atn. asp.
5. These rating systems follow the examples set in the Africa Growth and Opportunity Act (AGOA), which by 2003 applied to 39 countries; the remaining thirteen African states were vetoed by the White House for various reasons. AGOA conditionalities include adopting neoliberal policies, privatising state assets, removing subsidies and price controls, ending incentives for local companies and endorsing US foreign policy.
6. See http://www.opensociety.org.
7. According to Sam Olukoya:

   > Reparations are a crucial issue in the struggle for environmental justice in Nigeria. Many of the ethnic groups in the Niger Delta have drawn up various demands. A key document is the Ogoni Bill of Rights, which seeks reparations from Shell for environmental pollution, devastation and ecological degradation of the Ogoni area. Shell's abuses in Ogoniland were made infamous by the late playwright and activist Ken Saro-Wiwa, who was executed by the Nigerian government. (Olukoya 2001)
8. Subtract crime and family breakdown; add household and volunteer work; correct for income distribution (rewarding equality); subtract resource depletion; subtract pollution; subtract long-term environmental damage (climate change, nuclear waste generation); add opportunities for increased leisure time; factor in lifespan of consumer durables and public infrastructure and subtract vulnerability upon foreign assets.
9. Martinez-Alier elaborates with examples of ecological debt that are never factored into standard trade and investment regimes:

   nutrients in exports including virtual water . . . the oil and minerals no longer available, the biodiversity destroyed. This is a difficult figure to compute, for several reasons. Figures on the reserves, estimation of the technological obsolescence because of substitution, and a decision on the rate of discount are needed in the case of minerals or oil. For biodiversity, knowledge of what is being destroyed would be needed. (2003)

   Some of these cases are considered in the discussion earlier concerning depletion of natural resources. See also http://www.deudaecologica.org.

PART 4

CIVIL SOCIETY STRATEGIES FOR GENUINE CLIMATE JUSTICE

9

Oil, Climate Change and Resistance from the South

Joan Martinez-Alier and Leah Temper

There are new voices from the South clamouring to make themselves heard on climate change. They demand climate justice and refuse the alms offered by the North in the form of so-called flexibility mechanisms and adaptation loans, which transform the principle of 'the polluter pays' into 'the polluted adapt'.

The Kyoto Protocol has failed. Despite many admonitions from the Intergovernmental Panel on Climate Change (IPCC), the reality is that emissions of carbon dioxide (CO_2) in the world are going up by over 3 per cent per year. This is the failure of the countries that signed the Protocol and even more so, of those like the United States, who stayed outside the timid Kyoto Protocol framework and also of those not included in Annex I of the Rio de Janeiro climate change treaty of 1992 (the United Nations Framework Convention on Climate Change [UNFCCC]).

The world is currently burning about 85 million barrels of oil per day. As we approach peak oil (at 90 million barrels of oil per day? 100 million barrels of oil per day?), the price of oil goes up and up, despite efforts to get more of it by means foul or fair as in Iraq, the Niger Delta, the Amazon and other commodity frontiers. In Canada, rocketing oil prices have finally made exploitation of the Alberta oil sands profitable, whereby one barrel of oil equivalent is needed to produce three to five barrels of oil. The Hubbert curve is named after the geologist who 60 years ago predicted that the US peak oil would take place in the early 1970s. The road down the Hubbert curve will be terrible. Downhill will be harder than uphill.

The price of oil is followed by the price of natural gas. There are also conflicts around the world on gas extraction, like in Bolivia a few years ago where the gas contracts cost some dozens of human lives and a change of president, and in Mynamar where Unocal infringed on human rights when building a gas pipeline to Thailand. The juggernaut goes on, trampling indigenous peoples and biodiversity under its wheels.

Peak oil

Oil and gas prices are still too cheap, in the sense that neither local damages (externalities), nor their effects on climate change, are included. But their prices are rising rapidly because peak oil and later peak gas are fast approaching. Peak oil refers to the maximum output produced annually in the world, after which time extraction will decrease, while potential demand will still rise. Another effect of peak oil is that more and more energy is needed to pump the remaining oil out of the ground, what is referred to as energy return on energy investment (EROI.)

In contrast, the supplies of coal are plentiful and there is no Organisation of Petroleum Exporting Countries (OPEC) for coal to restrict supply. Therefore the first half of the twenty-first century is likely to be an era of coal. In the twentieth century the use of coal increased by a factor of six. The world's fastest growing economies, China and India, are fuelling their industrialisation with cheap, readily available coal, counteracting reductions in energy intensity elsewhere.

The trouble is that coal is socially and environmentally a very dirty business, whether procured by underground or open cast mining. Coal usually contains sulphur, which causes acid rain. Per unit of energy delivered, coal produces considerably more CO_2 than oil or gas. Nevertheless, the fear of emissions caps has not deterred a boom in the construction of coal-fired power plants, even in progressive European countries, with Romanian and Bulgarian massive coal deposits a growing supply source.

While the technology for capturing some carbon emissions from coal and storing them underground is becoming available at a cost, current prices for carbon emissions on the European market provide no incentive to do so. It is simply cheaper to pollute now and pay it off later. Or rather to pollute in Europe and then invest in a tree plantation or some other Clean Development Mechanism (CDM) project in the developing world. Because of this trend, human-produced CO_2 emissions in the world keep increasing in a trajectory that means that a concentration of 450 parts per million would be reached in little more than 30 years, while according to the IPCC, emissions should come down by 60 per cent in the next few decades.

A history of climate change

The intellectual history of the enhanced greenhouse effect is not yet common knowledge. It begins at least over one hundred years ago, when Svante Arrhenius from Sweden, a Nobel Prize winner in chemistry, published some calculations on the effects on temperature of a doubling or tripling of the contents of CO_2 in the atmosphere, with results very close to the present ones.

In 1938, electrical engineer G.C. Callendar published an article explaining that the combustion of coal would produce a slight increase in temperatures around the globe. According to him, there was nothing to worry about. Everybody knew that burning coal was good for the economy and human well-being and the increase in temperature was also good because it would extend the margin of cultivation to the North.

Twenty years later, at the end of the 1950s, Roger Revelle (who is featured in Al Gore's film, *An Inconvenient Truth*) and other scientists sounded a cry of alarm. Systematic measurements of CO_2 concentrations in the atmosphere were made. In the late 1980s, the IPCC began to get going.

This intellectual history is interesting for its own sake (late lessons from early warnings) but also because it bears on the historical responsibility for climate change that falls on the industrial counties. Should the question of responsibility go back to 1992 and the Rio de Janeiro Earth Summit treaty? Should it go back to 1960, or even further back?

An OPEC eco-tax?

The newly industrialised countries such as China and India do not want to talk about climate change. When they do, they argue that they should have the same opportunity to grow as the West did. On their side, the oil-exporting countries (as well as the coal-exporting countries, such as Colombia), do not want to hear about the enhanced greenhouse effect. The curb or the cap on emissions, if it ever came, would mean lowering the demand for fossil fuels.

Even in 1992 Saudi Arabia started to complain that it would claim compensation against those who were ready to spoil the oil market by unproven alarms about climate change. Today the official position of OPEC remains the same, only slightly modified of late by proposing to constitute a fund to subsidise research on carbon 'sequestration' technologies.

Therefore, it is all the more remarkable that at the OPEC meeting in Riyadh on 18 November 2007, President Rafael Correa of Ecuador, aware of a 2001 speech by Herman Daly in Vienna to the cartel leaders, proposed a new eco-tax on oil exports by OPEC countries with the explicit aim of lowering a little the demand for oil in order to diminish CO_2 emissions. The proceeds from the tax (the Daly-Correa tax?) should go for poverty reduction (including energy poverty reduction) and for alternative energies (meaning geothermal, wind and solar, and not, let us hope, agro-fuels or civil-military nuclear proliferation). Correa stated that to start with, the tax could be 3 per cent of the price of oil.

There is in this proposal an element of economic justice (many rich countries put heavy taxes on imported oil and gas, against the exporting countries). There is also an element of climate justice, based on a new awareness (among at least one of the smallest OPEC members) of the realities of the enhanced greenhouse effect and the international distribution of its causes and effects. Such realities are apparent in Ecuador with glaciers in the Andes losing ice cover and future sea-rise threatening Guayaquil.

Ecological debts

Other voices from the South asked in Bali for recognition of the ecological debts or the environmental liabilities owed from North to South. There is a public and a private aspect to this.

Firstly, countries that historically have produced and continue to produce more CO_2 per capita than the rest have a carbon debt. Jyoti Parikh, a previous member of the UN IPCC argued in 1995 that the average global emissions were about one ton per person per year. Industrialised countries produced three-quarters of these emissions, instead of the one-quarter that would have corresponded to them on the basis of population.

The difference was 50 per cent of total emissions, some 3 000 million tonnes at the time. Contemplating the increasing marginal cost of reduction, the first 1 000 million tonnes may perhaps be reduced at a cost of, say, US$15 per ton, but then the cost would increase very much. Taking an average of US$25 then, a total annual subsidy of US$75 billion was forthcoming from South to North (Parikh 1995).

The North has occupied the sinks (such as the oceans) and the atmosphere as a temporary deposit. The Northern countries are debtors and they should pay, as Anil Agarwal and Sunita Narain from the Centre for Science and Environment of Delhi argued as early as in 1991, basing their case in equal per capita emissions allowances (with populations of 1990).

Oil company liabilities

From the point of view of corporate accountability, many oil companies have done terrible damage to the local inhabitants and to other forms of life in the name of profit. In the Niger Delta, with high population density and an ecosystem of mangroves and agriculture, Ogoni and Ijaw activists have often pointed out the inconsistency between the international rhetoric on saving the world's climate and the local reality of oil extraction, waste dumping and gas burning at the cost of so many human lives.

Shell has not been held accountable either for environmental and social damage or for the death of Ken Saro-Wiwa and his comrades in 1995. At present Shell continues to flare gas despite new laws prohibiting the practice. Since almost 30 years ago the government of Nigeria has been setting deadlines for oil companies to stop gas flaring. The current deadline to stop gas flaring was 1 January 2008. But the oil companies – who save money by gas flaring, thus increasing their unpaid environmental and social liabilities – want at this very moment to postpone the deadline yet again to 2011.

As defined by the non-governmental organisation (NGO) Environmental Rights Action (ERA) whose president is the writer and activist, Nnimmo Bassey, gas flaring (that also takes place in the Amazonia of Ecuador or Peru where oil is extracted and in so many other places) is the burning off of gas, which sends a cocktail of poisons into the atmosphere. In the mix are CO_2 and methane, which are both major causes of global warming. Gas flaring causes acid rain, which acidifies the lakes and streams and damages crops and vegetation. It reduces farm yields and affects human health, lives and livelihoods. Gas flaring increases the risk of respiratory illnesses, asthma and cancer. It often causes painful breathing, chronic bronchitis, decreased lung function, itching, blindness, impotency, miscarriages and premature deaths.

As reported by ERA, since 1979, the oil companies have simply ignored government deadlines and court orders to end gas flaring. In a suit brought by the Iwerekhan community, the judge ruled that gas flaring 'is a gross violation of their fundamental right to life, including healthy environment and dignity of the human person'.

Again, Ecuador provides other lessons. The court proceedings against Texaco (now Chevron-Texaco) that started in New York under the Alien Tort Claims Act in 1993 are now reaching a conclusion in a court in Lago Agrio, an oil-polluted township in the Sucumbios province. There might be an agreement out of court. The damages (because of oil spills, gas flaring, over 600 pools of polluted extraction water and the resultant cancer cases, extinct tribes and lost biodiversity) are now being quantified in money terms because this is the nature of the court case (a civil suit for damages and not a criminal case).

The damages caused by Texaco between 1970 and 1990 (when it extracted over 1.3 billion barrels of oil) must now be valued in money terms. It would seem that in a forensic context, there is no room for debates on incommensurability of values. There are damages in terms of lost human health, the destruction of local indigenous groups, soil and water polluted and loss of biodiversity. Texaco made a conscious decision not to reinject the water, standard practice in the United States at the time, nor to line the waste pits. These damages could be estimated in terms of saved costs, or in terms of the economic value of human suffering and nature spoiled.

The Frente de Defensa of Amazonia based in Lago Agrio is headed by the resilient leader Luis Yanza and the young local lawyer Pablo Fajardo, who in December 2007 has obtained a CNN award for 'unknown world protagonists'. They are fighting an unequal fight against Chevron-Texaco, although they have been able to mobilise some resources: pictures of the singer Sting and his wife, Al Gore, Luis Yanza and Pablo Fajardo, all of them crossing arms and smiling together taking the victims' side in the Chevron-Texaco case in Ecuador have appeared in the US media.

A claim for about US$6 billion has often been mentioned. The present value of this sum (at a rate of interest of only 5 per cent, and taking also into account the loss of purchasing power of the dollar in the last 20 or 30 years) would exceed US$20 billion.

The lesson from Lago Agrio is that oil, coal and gas companies can no longer get away with not paying for their social and environmental liabilities, even when they are operating in places where human life is cheap and the destruction of nature is not carried into the bottom line of the profit-and-loss account. Since 1993, it has been civil society, through its organisations and support groups in Ecuador and abroad, which has pushed the case.

The Yasuni ITT Proposal

Finally, another innovative oil policy coming from civil society is the Ishpingo-Tambococha-Tiputini (ITT) Yasuni proposal, also in Ecuador (Martinez-Alier 2007). The idea was first expressed in the OilWatch position paper in Kyoto in 1997, arising from Ecuador and in the Niger Delta (through the network OilWatch): stopping oil extraction and gas flaring was a contribution to the struggle against climate change. The consumers of oil, the governments of oil-exporting or oil-importing countries, should recognise that civil society organisations fighting for environmental justice are more advanced than they are.

Thus, in the ITT field in the Yasuni National Park, about 920 million barrels of heavy oil would remain in the ground in perpetuity or in a moratorium indefinitely, in an area that holds unique biodiversity and which is inhabited by indigenous groups, some living in voluntary isolation. An ancillary benefit of keeping this oil in the ground (apart from respecting nature and human rights) is that the CO_2 that would be produced when burning the oil elsewhere is 'repressed' underground. The avoided emissions of CO_2 are of the order of 410 million tonnes from the oil, plus some more from the avoided gas flaring and avoided deforestation. Ecuador is asking for a part of the money from outside in recognition of its foregone monetary revenue. The Frente de Defensa of Amazonia, if successful in the Chevron-Texaco case, would probably be willing to make a substantial financial contribution to keeping oil in the ground in the Yasuni ITT field.

At present, there is strong support inside the government of Ecuador for this project, which was launched by the then minister of energy, Alberto Acosta, early in 2007. Acosta is now the president of the assembly that is writing the new Constitution: he has stated that the ITT field and other natural parks should be declared out of bounds for the oil industry. However, support from President Rafael Correa is not firm because he is by training a development economist with anti-environmental inclinations.

This project, if successful, could be copied elsewhere – for instance in U'Wa territory in Colombia whose inhabitants have argued against the oil companies that the land is sacred, in the Niger Delta, or in some of the worst coal mines in the world in India or China.

10

Beyond Bali

Brian Tokar

With all the fanfare that usually accompanies such gatherings, delegates to the December 2007 United Nations (UN) climate talks on the Indonesian island of Bali returned to their home countries declaring victory. Despite the continued obstructionism of the US delegation, the negotiators reached a mild consensus for continued negotiations on reducing emissions of greenhouse gases and, at the very last moment, were able to cajole and pressure the United States to sign on.

But in the end, the so-called 'Bali roadmap' added little besides a vague timetable to the plans for renewed global climate talks that came out of a similar meeting two years previously in Montreal. With support from Canada, Japan and Russia and the acquiescence of former ally Australia, the US delegation deleted all references (except in a non-binding footnote) to the overwhelming consensus that reductions of 25 to 40 per cent in annual greenhouse gas emissions are necessary by 2020 to forestall catastrophic and irreversible alterations in the Earth's climate.

In Kyoto in 1997, then Vice-President Al Gore was credited with breaking the first such deadlock in climate negotiations: he promised the assembled delegates that the United States would support mandatory emissions reductions if the targeted cuts were reduced by more than half and if their implementation were based on a scheme of market-based trading of emissions. The concept of 'marketable rights to pollute' had been in wide circulation in the United States for nearly a decade, but this was the first time a so-called 'cap-and-trade' scheme was to be implemented on a global scale. The result, a decade later, is the development of what British columnist George Monbiot has aptly termed 'an exuberant market in fake emissions cuts' (2007b). Of course, the United States never signed the Kyoto Protocol and the rest of the world has had to bear the consequences of managing an increasingly cumbersome and ineffectual carbon trading system.

Given the increasingly narrow focus on carbon trading and offsets as the primary official response to global climate disruptions, it is no surprise that Bali resembled, in the words of one participant, 'a giant shopping extravaganza, marketing the Earth, the sky and the rights of the poor'. All manner of carbon brokers, technology developers and national governments were out displaying their wares to the thousands of assembled delegates and non-governmental organisation (NGO) representatives. Numerous international organisations used the occasion of Bali to release their latest research on various aspects of global warming, including an important new report from the Global Forest Coalition highlighting the consequences for the world's forests of the current global push to develop so-called 'biofuels' from agricultural crops, grasses and trees.

Indeed, the problem of deforestation, which is now responsible for 20 per cent of worldwide carbon dioxide (CO_2) emissions, was very much on the agenda in Bali. In anticipation of a future UN scheme to address what it calls 'Reducing Emissions from Deforestation and Degradation' (REDD), the World Bank announced the creation of a new Forest Carbon Partnership Facility (FCPF). World Bank funds will now be available for governments seeking to preserve forests, but given the Bank's long history of funding environmental destruction, observers remain sceptical. The effort mainly perpetuates the fatuous idea that wealthy nations (and individuals) can offset their excessive CO_2 emissions by paying for nominally carbon-saving projects in poorer countries. Carbon offsets have already spurred the replacement of vast native forests with timber plantations, more readily assessed for their carbon sequestration potential and able to be harvested for energy crops, such as palm oil and highly speculative cellulose-derived ethanol.

A statement issued by nearly 50 critical NGOs assembled in Bali stated, in part, 'The proposed REDD policies could trigger further displacement, conflict and violence; as forests themselves increase in value they are declared off limits to communities that live in them or depend on them for their livelihoods'.

A central underlying assumption of REDD, as with similar World Bank initiatives in recent years, is that traditional forest-dwelling communities are incapable of managing their forests appropriately and that only international experts affiliated with the Bank, national governments and compliant environmental organisations such as Conservation International and the Worldwide Fund for Nature (WWF) are capable of doing so. Ultimately, timber companies and plantation managers, in league with the World Bank, will be demanding, in the words of Simone Lovera of the Global Forest Coalition, 'compensation for every tree they don't cut down'.

The Bali meetings also led to the creation of a new UN fund to help poor countries adapt to climate changes. The Intergovernmental Panel on Climate Change (IPCC) made it clear in its exhaustive 2007 report that the people least responsible for climate change will likely bear the worst consequences, as they are most vulnerable to the widespread increases in floods, droughts, wildfires and other effects of a rapidly changing climate. The UN's biannual Human Development Report, also released in Bali, states that at least one out of every nineteen people in the so-called developing world was already affected by a climate-related disaster between 2000 and 2004.

The new UN adaptation fund will be managed by the Global Environmental Facility (GEF), a semi-independent partnership of the UN's environment and development programmes and the World Bank and funded through a 2 per cent levy on carbon offset transactions under the Kyoto Protocol's Clean Development Mechanism (CDM). The CDM's carbon offset schemes, however, have been widely criticised for manipulations, abuses and the funding of highly questionable projects including, once again, large scale commercial timber plantations displacing tropical rainforests. The new adaptation fund binds governments of poor countries even more tightly to the questionable practice of carbon offsets, even as it offers only a miniscule fraction of the estimated US$86 billion needed just to sustain current UN poverty reduction programmes in the face of the myriad new threats related to climate change.

So while the continued obstructionism of the Bush administration is the main story in the international press, the successful entrenchment in the UN system of 'market-driven' policies introduced by the Clinton-Gore administration may prove to be the more lasting obstacle to real progress on global warming. Carbon trading and offsets help to further enrich Gore's colleagues in the investment banking world, but contribute almost nothing to actually reducing emissions of CO_2 and other greenhouse gases. What are we to do?

Activists across the United States and in other industrial countries have begun to dramatise the reality of potentially catastrophic global warming and pressure their governments to do something about it. Al Gore's *An Inconvenient Truth* has had a positive educational impact, as has the latest IPCC report, documenting the 'unequivocal' evidence that global warming is real and that we can already see the consequences. But most public events up to now, at least in the United States, have been rather timid in their outlook and minimal in their expectations for real changes. The failure of the Bali talks suggests the urgency of a far more pointed and militant approach, a genuine People's Agenda for Climate Justice. Such an agenda would have at least four central elements:

- Highlight the social justice implications of global climate disruptions. Global warming is not just a scientific issue and it's certainly not mainly about polar bears. As the UN's

Human Development Report describes so eloquently, global warming is a global justice issue and its implications for the half of the world's people that live on less than US$2 per day are truly staggering. Bringing home these implications can go a long way toward humanising the problem and raise the urgency of global action.

- Dramatise the links between US climate and energy policies and US military adventures, particularly the war in Iraq, which is without question the most grotesquely energy-wasting activity on the planet today. Author Michael Klare has documented that troops in the Persian Gulf region consume 3.5 million gallons of oil a day and that worldwide consumption by the US military – about four times as much – is equal to the total national consumption of Switzerland or Sweden. In October 2007, people gathered under the banner of 'No War, No Warming' blocked the entrances to a Congressional office building in Washington, demanding an end to the war and real steps to prevent more catastrophic climate changes. Similar actions across the country could go a long way toward raising the pressure on politicians who consistently say the right thing and blithely vote the opposite way.
- Expose the numerous false solutions to global warming promoted by the world's elites. Billions of dollars in public and private funds are wasted on such schemes as a revival of nuclear power, mythical 'clean coal' technologies and the massive expansion of so-called biofuels (more appropriately termed agrofuels): liquid fuels obtained from food crops, grasses and trees. Carbon trading and offsets are described as the only politically expedient way to reduce emissions, but they are structurally incapable of doing so. We need mandated emission reductions, a tax on CO_2 pollution, requirements to reorient utility and transportation policies, public funds for solar and wind energy and large reductions in consumption throughout the industrialised world. Buying more green products won't do; we need to buy less!
- Envision a new, lower-consumption world of decentralised, clean energy and politically empowered communities. Like the anti-nuclear activists of 30 years ago, who halted the first wave of nuclear power in the United States, while articulating an inspiring vision of directly democratic, solar-powered communities, we again need to dramatise the positive, even utopian, possibilities for a post-petroleum, post-mega-mall world. The reality of global warming is too urgent and the outlook far too bleak, to settle for status quo false solutions that only appear to address the problem. The technologies already exist for a locally controlled, solar-based alternative, at the same time that dissatisfaction with today's high consumption, high debt 'American way of life' appears to be at an all-time high. Small experiments in living more locally, while improving the quality of life, are thriving everywhere, as are experiments in community-controlled

renewable energy production.[1] Al Gore is correct when he says that political will is the main obstacle to addressing global warming, but we also need to be able to look beyond the status quo and struggle for a different kind of world.

Note

1. See http://www.solartopia.org.

11

Conclusion

Leave the Oil in the Soil

Patrick Bond

In October 2004, the Durban Declaration on Carbon Trading was drafted by environmental justice organisations, community groups and concerned citizens who spent the prior week carefully analysing the climate mitigation potential of the emissions market, before rejecting the strategy (see Appendix 2 in this volume). As argued in this book, Durban was an appropriate place to launch our movement, on the basis of powerful traditions of anti-corporate environmental activism and in particular the struggle by Sajida Khan to close the Bisasar Road dump, foiled by the municipality's hope for a Clean Development Mechanism (CDM) award of US$15 million to keep it open.

State of the climate

Since then, yet more evidence of global warming has emerged. Leading officials concede that the brutal hurricanes of September 2005 were mainly attributable to higher Gulf of Mexico water temperatures. By January 2006, even *Fortune* magazine could run this hair-raising story:

> Like the tourists on Phuket beaches who stood and gazed at an oncoming tsunami because it was outside their experience, society is reacting to the coming wave of climate change without urgency. People still believe that the science is controversial and the threat of climate change far off in the future; and while a few businesses, notably major insurers, have begun to adapt, governments are responding only slowly, as the lack of progress at this fall's international forum in Montreal showed.
>
> The wave is coming, though. The last decades of the 20th century saw an unmistakable and extraordinary warming. During this same period, we

> suffered by some measures the strongest El Niño in 130 000 years and a swarm of statistically extraordinary droughts, floods, and other weather extremes . . .
>
> The Earth's heat-distribution system has already begun shifting massively in response to rising levels of greenhouse gases. Precipitation patterns, the change of seasons, storm intensity, sea ice, glaciers, temperatures on the tundras – all are in flux. As scientists nervously monitor sea and air currents for signs of major shifts, many believe that today's proliferation of weather extremes may be the prelude to another epochal transition, a possibility first flagged by the great oceanographer Wallace Broecker in the journal *Science* in 1997.
>
> How bad could it get? Imagine Europe suffering floods and heat waves on a vastly greater scale than those endured in 2002 and 2003, while northern regions experience intermittent deep freezes as atmospheric and ocean circulations struggle to find new equilibrium. At the same time, droughts and floods not seen since ancient times would afflict some of the most densely populated regions on earth. (Linden 2006)

State of the climate mitigation debate

A scientific consensus now appears unshakable: by 2050, the world requires 80 per cent reductions in CO_2 emissions to prevent tipping the world into an unmanageable environment and potentially a species-threatening crisis. Yet the options being contemplated in global and national public policy debates to take us to 80 per cent reductions are nowhere near what is required, for several reasons.

The first is that the global balance of forces appears adverse to the kinds of radical changes required. As a mid-2008 report from Bonn puts it:

> Another round of talks on the road towards a new global deal on climate change was wrapping up in Germany on Friday, battered by criticism that progress had been negligible. The 12-day haggle under the 192-nation United Nations Framework Convention on Climate Change (UNFCCC) was the second since the accord in Bali, Indonesia, last December that set down a 'road map' towards a new planetary treaty . . . India representative Chandrashekar Dasgupta deplored 'the lack of any real progress' in Bonn and 'a deafening silence' among industrialised countries, save the European Union. (Agence France Press 2008)

This failure was true to form, as the very people and corporations responsible for these problems – especially in the US/EU-centred petro-mineral-military complex and associated financial agencies such as the World Bank – renewed their grip on power in the late 2000s. Here are some of their 'accomplishments':

- With only minor embarrassment, the Bush regime went to the Bali Conference of Parties (COP) in December 2007 and once again held back negotiations on a climate regime, which should have set measurable targets and established firm accountability systems, but did neither.
- Without hesitation, president of the World Bank, Robert Zoellick advanced his institution as a vehicle for massive financing of climate mitigation and adaptation via the Global Environmental Facility (GEF), in spite of the Bank's leading role in fossil-fuel extraction, as well as an official statement regarding 'the negative experiences that most developing countries have with their interaction with the GEF and the World Bank in respect of the funding of climate related projects'. Such a statement – by Marthinus van Schalkwyk (2007c: 1), reflected the South African government's desire to rehabilitate the Bank, but via a necessary acknowledgement of its failures, so as to get Southern buy-in.
- Without any sense of urgency, in the November 2006 climate change negotiations in Nairobi, the major powers ignored the British government's Stern Review on the cost of climate change and need for emissions reductions, endorsed increased carbon trade for Africa, and began developing dubious strategies such as monocultural timber plantations, biofuels and genetic engineering technology through the CDM. Meanwhile the South African government is playing a facilitating role in global warming, in cheap electricity deals with the Canadian aluminium company Alcan and a co-operation agreement on climate policy with the reactionary Australian government.
- Without concern for future generations, the G8 countries refused to reverse course at St Petersburg in July 2006, or earlier at the Montreal meeting of the Kyoto COP in December 2005.
- Without understanding, key players in the carbon market – especially EU bureaucrats responsible for generously dishing out historic rights to pollute through their emissions trading scheme – witnessed the price of carbon plummet by half in May 2006, as fictional supply of credits soared above fictional demand.
- Without shame, the largest oil corporations visited Johannesburg for the World Petroleum Congress in September 2005 to celebrate their world-historic profits.
- Without a worry for his legitimacy, George W. Bush established a new alliance of hyper-polluters – the United States, Australia, India and China – in July 2005 to again foil serious carbon reduction efforts.

- Without caveat, the G8 leaders met in Gleneagles in July 2005, giving the architect of the war in Iraq, then World Bank president, Paul Wolfowitz, the green light to accelerate his institution's prolific contribution to climate change.
- Without a thought to Wolfowitz's legacy or agenda, the then chair of the World Bank/ International Monetary Fund (IMF) Development Committee, South African finance minister Trevor Manuel, welcomed him to his new job in April 2005, calling him 'a wonderful individual . . . perfectly capable'.

Within the gridlock represented by these failures, two strategies to combat climate change can be discerned: emissions cap-and-trade options and carbon taxation; outside the gridlock are two others: command and control of emissions, and alternative grass-roots climate change mitigation strategies. The latter two are what we insist will be necessary to save the planet, yet it is the former two strategies that are still ascendant.

The current state of debate, in mid-2008, divides those who would want the world economy to slowly and painlessly adapt to CO_2 abatement strategies, and those who would advocate dramatic emissions cuts in a manner that is both redistributive (from rich to poor and North to South, and in the process male to female), and sufficiently shocking to economic structures and markets that major transformations in production and consumption are compelled.

There are some who argue that, along this spectrum, market-based instruments – either a cap-and-trade system or carbon tax (or some hybrid) – will have the capacity to rope in the major CO_2 emitters and compel them to reduce greenhouse gases as an economic strategy. A debate has emerged about how to make mitigation more efficient. As the US Congressional Budget Office explains:

> The most efficient approaches to reducing emissions of CO_2 involve giving businesses and households an economic incentive for such reductions. Such an incentive could be provided in various ways, including a tax on emissions, a cap on the total annual level of emissions combined with a system of tradable emission allowances, or a modified cap-and-trade program that includes features to constrain the cost of emission reductions that would be undertaken in an effort to meet the cap. (2008)

To retiterate the definitions here, the 'cap' means that each major point source of emissions – usually in the form of a country or firms within a country – would be granted an emissions permit for each ton of CO_2 released into the atmosphere. The cap would gradually reduce to the point that by 2050, the 80 per cent target would be met. The crucial point is that through the trade, flexibility can be attained, so as to achieve more efficient greenhouse gas reduction. Those with the opportunity to make bigger cuts should do so and sell their 'hot air' – the emissions saved above and beyond what is required at any given point in time – to those who have a harder time making the required cuts. Such a trading strategy would keep the high-emissions businesses alive until they have time to adapt. Auctioning the permits would give governments a dependable revenue stream, which could be used to invest in renewable energy and other innovations. In the United States, US$300 billion per year is anticipated as feasible income (at US$10–15 per metric ton of CO_2) by reducing emissions 80 per cent below 1990 levels by 2050.

Another version of a market-based climate change mitigation system – which either enforces underlying economic dynamics or changes them – is a tax on greenhouse gas emissions. Such a tax would take the production system as given and alter the demand structure. According to an assessment by the US Congressional Budget Office:

> A tax on emissions would be the most efficient incentive-based option for reducing emissions and could be relatively easy to implement. If it was coordinated among major emitting countries, it would help minimize the cost of achieving a global target for emissions by providing consistent incentives for reducing emissions around the world. If other major nations used cap-and-trade programs rather than taxes on emissions, a U.S. tax could still provide roughly comparable incentives for emission reductions if the tax rate each year was set to equal the expected price of allowances under those programs. (2008)

The major problems with taxation are tax-avoidance capacities of influential industries and incidence: who will pay the bill? As noted below, there are ways to design a tax system with a strongly redistributive outcome and in the process to incentivise transformative economic strategies. However, a dramatic shift in political power is required for such an outcome.

A more equitable version of emissions trading advocacy comes from those who recommend a per-capita strategy, oriented to social justice along North-South lines, combined with trading. The per-capita right-to-emit has been advocated through contraction-and-convergence and greenhouse development rights strategies.

The alternatives to such market-based strategies typically fall into state-oriented command-and-control and activist 'direct action'. The rationale here is, typically, that the application of market incentives – and in the process, the granting of pollution rights – cannot generate the cuts needed to save the species from severe damage due to climate change. Instead, a variety of strategies and tactics that would explicitly cut greenhouse gas emissions is preferable. Some of the strategies – a switch to renewable energy, changed consumption patterns, new production and consumption incentives through punitive taxation and 'keep the oil in the soil and the coal in the hole' campaigns – are already being adopted by some activists.

In mid-2008, the most important single site of debate was the US Congress, where a cap-and-trade law proposed by Senators Joe Lieberman and John Warner was narrowly defeated. Although there were two committed US presidential candidates in the November 2008 election with aggressive positions on climate change – Ralph Nader (Independent) and Cynthia McKinney (Greens) – they did not stand a chance of making this a major issue, or making a dent in the outcome. The two candidates who could set the US climate agenda from 2009 onwards, Barack Obama and John McCain, both support the cap-and-trade concept. The primary difference is that Obama supports cap and auction, while McCain would give out emissions permits to large CO_2 polluters for free, at least initially.

The Environmental Defence Fund argued that core support for cap and trade in the US Congress might represent an opportunity in 2009 for a major legislative initiative. However, opposition to Lieberman-Warner by environmentalists and other progressive organisations – including Greenpeace, Friends of the Earth, MoveOn.org, CREDO Mobile and Public Citizen – was a result of its inclusion of support for nuclear energy, its inadequate emissions cap, the adverse impact on low-income people and other problems inherent in carbon trading. Environmental justice organisations lobbied not for cap and trade, but for a robust and fair carbon tax instead.

The other main site of debate is Europe, whose Emissions Trading Scheme (ETS) has been hotly contested. Due to the large reliance upon controversial offsets, as well as the ETS price crash in April 2006 once a flood of emissions permits were released to companies on a gift (non-auctioned) basis, there remains doubt about the ability of the ETS authority to tackle the challenge of regulating emissions. Moreover, roughly €50 billion worth of pollution rights (measured at €30 per ton) were being transferred to large European CO_2 emitters annually through the ETS. According to Jutta Kill (2008), five lessons emerged from the ETS experience:

1. Over-allocation of permits due to intensive industry lobbying during the allocation process led to price collapse of ETS permit prices in April 2006 and few permit trades for compliance purposes. Similar price collapse due to over-allocation has been reported for the New South Wales emissions trading scheme. Lack of a stringent cap has undermined the emissions trading scheme.

Slight tightening of the cap for the second phase of the ETS from 2008–2012 in the wake of the failure and price collapse during Phase 1 has been offset by increasing the hole in the cap: across the board, companies are allowed to use significantly more offset credits from CDM and joint implementation (JI) projects during Phase 2 compared to Phase 1 of the ETS. Several reports have shown that the shortfall of permits resulting from the tightening of the cap in Phase 2 will be filled to 88–100 per cent by increased volume of offset credit influx into the ETS.

Transfers of wealth to polluters by EU countries.

Privatisation of atmospheric world carbon dump by the EU ETS	Phase 1: Gift to big business (MT CO2)	2005 emissions	Phase 2: Approved gift to big business	Increase decrease in gift to	Gift = x% of 'world carbon dump' (IPCC)	Yearly value of gift @ €30/t
Czech R	97.6	82.5	86.8	+5%	-1–2%	€2.6b
France	156.5	131.3	132.8	+1%	-1–3%	€4.0b
Germany	499	474	453.1	-4%	-5–9%	€13.6b
Netherlands	95.3	80.4	85.8	+7%	-1–2%	€2.6b
Poland	239.1	203.1	208.5	+3%	-2–4%	€6.3b
Spain	174.4	182.9	152.3	-17%	-2–3%	€4.6
Sweden	22.9	19.3	22.8	+18%	<1%	€0.7b
UK	245.3	242.4	246.2	+2%	-3–5%	€7.4b
Total EU	1 815.7	1 672.5	1 650.7	-1%	-17–34%	€49.52b

Source: Kill 2008

2. Free allocation of emission permits has led to record windfall profits to energy utilities and some of the highest emitting industry sectors in the European Union. It seems that 100 per cent auctioning in the third phase of the ETS is increasingly considered as the only remedy to salvage the ETS. Capping emissions without 100 per cent auctioning selects against immediate investment in long-term structural change. Short-term and uncertain price signals discourage structural change and cost-spreading discourages innovation.
3. Any influx of offset credits into the ETS will undermine effectiveness due to risk of development of a 'lemons market' as a result of unverifiable quality of offset credits. This is of concern, particularly given the increasing evidence

that up to one-third of CDM projects either already registered or in the process of CDM registration are considered 'non-additional' by CDM experts.

4. There is increasing acknowledgement, including from the private sector, that emissions trading will not provide the incentives and price signals required to trigger significant investments and research and development into zero-carbon and low-carbon technologies, which is required to be able to achieve the emissions cuts required to avert climate chaos.
5. Increasing signs that more effective approaches to switch to zero-carbon economies are held back for fear of jeopardising the European Union's flagship ETS. A leaked government internal note from the United Kingdom, for example, reveals a deep concern that achieving the 20 per cent renewable energy target itself could present a 'major risk' to the European Union's ETS, for which London has become a major centre of exchange. Combined with the European Union's drive to greater energy efficiency, increasing the share of renewable energy could cause a carbon price collapse and make the ETS 'redundant', the note says.

A crucial determinant of the impact of market mechanisms, whether carbon trades or taxes, is the problem of our unreliable understanding of carbon price elasticity, i.e. what happens to the demand for carbon-related products when their price changes, either in small increments or dramatically. The latest data and their implications for environmental justice are reviewed below. In addition, a series of less publicised alternatives are in continual evolution, including the contraction-and-convergence and greenhouse development rights strategies for per-capita emissions rights, which also involve trading.

In contrast to market-related approaches, command-and-control strategies for emissions reductions have an important history. However, for public policy to evolve in a just and effective way on climate emissions, a much stronger set of measures will be required. These will mix the set of command-and-control strategies associated with prior emissions controls (e.g. chlorofluorocarbons in the 1996 Montreal Protocol and many European regulations of emissions) and the national state strategy known as 'leave the oil in the soil' (and 'leave the coal in the hole'), with direct grass-roots action against greenhouse gas emission points (such as coal facilities).

Reformist and non-reformist reforms

The most important lesson is in the overall failure of market strategies to date. There are intrinsic, deep-level problems in the new emissions markets, both on their own terms and with respect to the climate and peoples most vulnerable. What is required is agreement on the strategic orientation and the kinds of alliances that can move the debate forward. In terms of the debate over market solutions to the climate crisis, consider the late French sociologist Andre Gorz's distinction in his books *Strategy for Labour* (1967) and *Socialism and Revolution* (1973) between 'reformist reforms' and 'non-reformist reforms':

1. *Reformist* reforms undergird, strengthen and relegitimise the main institutions and dynamics in the system that cause the climate change problem, and thus weaken and demobilise environmental and social justice advocacy communities through co-option.
2. *Non-reformist* reforms undermine, weaken and delegitimise the climate change system's main institutions and dynamics, and consequently strengthen its critics, giving them momentum and further reason to mobilise.

The discussion so far allows us to distinguish four market-based emissions mitigation initiatives along this spectrum:

1. *carbon trades without auctions* where pollution permits are grandfathered in, as in the European ETS, are now so widely delegitimised, that only US Republican Party candidate John McCain supports them;
2. *carbon trades with auctions* will increasingly dominate discussions, especially in the United States if Barack Obama is elected president in November 2008, in part because they have the support of many mainstream commentators and large environmental organisations;
3. *carbon taxes* either aimed to be revenue-neutral, or to raise funds for renewables and socio-economic transformation, will continue to be seen as the main progressive alternative to carbon trading, even though such taxes do not address more fundamental power relations or achieve systematic change required to avert climate disaster; and
4. *greenhouse development rights, contraction-and-convergence* and other *per-capita 'right to pollute' strategies* with a North-South redistributive orientation are also advocated by eloquent environmentalists and some Third World leaders, and entail a trading component and the property right to emit.

Each strategy has major disadvantages by virtue of being located within market-based systems, especially during a period of extreme financial volatility, during which energy-related securities (including emissions credits) have been amongst the most unreliable measures of value. As a result, we can conclude that the first two are reformist reforms, and the latter two have non-reformist *possibilities*. There are two further non-reformist alternatives – command-and-control emissions prohibitions and local supply-side strategies (a kind of command-and-control from below) – that bear consideration once the market-based strategies are briefly reviewed.

A central problem is that reformist reforms can be *counterproductive* to mitigating climate change. In short, it is possible that an exploitative system becomes even stronger in the wake of an eco-social change campaign. If campaigners unwittingly adopt the same logic of the system, and turn for change implementation to the kinds of institutions responsible for exploitative damage, and moreover also restore those institutions' credibility, the reforms may do more harm than good.

To illustrate, if mainstream environmentalists endorse World Bank strategies to commodify forests through the Reducing Emissions from Deforestation and Degradation (REDD) programme, their co-optation inevitably strengthens the Bank – responsible for vast climate damage as a major fossil-fuel investor – and weakens the work of indigenous people and environmental activists. The reformist-reform logic appears in the case of a Brazilian meat-packing plant in the Amazon that coincides with the Bank's investments in forest protection. There are, in such cases, persuasive advocates of reform, such as Dr Daniel Nepstad of Woods Hole Research Institute, who accept the basic parameters of the system's logic, namely the ongoing exploitation of the Amazon, and who seek to tame that process using World Bank resources:

> The irony is that at the same time the World Bank was launching the Forest Carbon Partnership Facility, the International Finance Corporation [a World Bank agency] was making a loan to the Bertin meat-packing plant in the Brazilian Amazon. The loan aims to set up a sustainable supply of beef for an ecological meat-packing facility in Marab in the state of Para. What upset the protestors was the idea that the same institution would be accelerating deforestation by expanding the capacity to process meat in the Amazon region as it creates this mechanism for compensating nations for reducing their emissions.
>
> Our own feeling on this is that there comes a point where we have to acknowledge that the region is undergoing an economic transformation and if we can find a powerful lever for commodifying how this transformation takes place – putting a premium on legal land-use practices, legal deforestation, the gradual elimination of the use of fire – we should take it. For me that trumps the negative consequences of setting up increased capacity in the

> region. In other words, I really do believe that there are many responsible cattle ranchers and soy farmers in the Amazon who are waiting for some sort of recognition through positive incentives.
>
> The incentive could be a very small mark up – literally a few cents per pound of beef sold – but it would send a signal to these ranchers that if they want to participate in the new beef economy, they better have their legal forest reserve in order or have compensated for it, maintain or be in the process of restoring their riparian zone forests, control erosion, and get their cows out of the streams and into artificial watering tanks. There is a whole range of positive things that can happen once cattle ranchers see that if they do things right they are rewarded. This means that as Brazil moves forward as the world's leading exporter of beef – with tremendous potential to expand – we have a way to shape that expansion as it takes place to reduce the negative ecological impacts. (cited in Butler 2008)

Such logic is also evident in efforts to reform carbon trading by advocating the auctioning of emissions permits. In opposition to reformist reforms, a coalition of 32 indigenous peoples (and environmental allies) lobbied against the REDD programme:

> Given the threat to Indigenous Peoples' Rights that REDD represents, we call on the United Nations Permanent Forum on Indigenous Issues to recommend strongly to the UNFCCC, the UN Forum of Forests, concerned UN agencies such as UNEP, the World Bank, the Special Rapporteur on Human Rights and Fundamental Freedoms of Indigenous Peoples and nation states that REDD not be considered as a strategy to combat Climate Change but, in fact, is in violation of the UN Declaration on Indigenous Peoples. Moreover, we also urge the Permanent Forum to recommend strongly to the Convention on Biological Diversity that the implementation of the programme of work on Forests and biodiversity prohibit REDD. We also further urge that Paragraph 5 be amended to remove 'clean development mechanism, the Clean Energy Investment Framework, and the Global Environment Facility'. These initiatives do not demonstrate good examples of partnership with indigenous peoples. There are many CDM projects that have human rights violations, lack of transparency and have failed to recognize the principles of Free, Prior and Informed Consent.

The statement was endorsed by an impressive range of indigenous people's organisations: Indigenous Environmental Network, CORE Manipur, Federation of Indigenous and Tribal Peoples in Asia, Na Koa Ikuiku Kalahui Hawaii, Indigenous World Association, CAPAJ-Parlamento del Pueblo Qollana, International Indian Treaty Council, Amazon Alliance, COICA, Instituto Indigena Brasileno para la Poropiedad Intelctual, The Haudenosaunee Delegation, Agence Kanak de Developpement, Mary Simat-MAWEED, Marcos Terena-Comite Intertribal-ITC-Brasil, Land is Life. ARPI-SC-Peru Amazonia, Asociaciones de Mujeres Waorani de la Amazonia AMWAE, Kus Kura S.C., Indigenous Network on Economic and Trade, Aguomon FEINE, Friends of the Earth International, Amerindian Peoples Association, FIMI North America, L. Ole L. Lengai-Sinyati Youth Alliance, Beverly Longid-Cordillera People's Alliance Philippines, Red de Mujeres Indigenas sobre Biodiversidad de Abgatala, Fundacion para la Promocion de Conocimiento Indigena, Asociacion Indigena Ambiental, INTI-Intercambio Nativa Tradicional Internacional, Global Forest Coalition, Fuerza de Mujeres Wayuu, and Caf' ek.

In contrast to reformist reform initiatives such as REDD, non-reformist reforms are generated by campaigns that explicitly reject the underlying logic of climate change, i.e. fossil-fuel exploitation. Such reforms legitimate the *opponents* of the system, not the system itself, and lead to further mobilisation, rather than to the movement's co-optation. An example is the partially successful struggle to 'keep the oil in the soil' in the Yasuni National Park waged for several years by the Quito non-governmental organisation (NGO) Acción Ecológia and its OilWatch allies. The campaign advanced rapidly in 2007, when Ecuadoran president, Rafael Correa, declared his intent to leave US$12 billion worth of oil reserves untouched in perpetuity, in exchange for anticipated payments from international sources – not as a carbon offset, but instead to be considered part of the North's repayment of its 'ecological debt' to the South. The aim of the proposal is to provide a creative solution for the threat posed by the extraction of crude oil in the Ishpingo-Tambococha-Tiputini (ITT) oil fields, which are located in the highly vulnerable area of Yasuni National Park. The proposal would contribute to preserving biodiversity, reducing carbon dioxide emissions and respecting the rights of indigenous peoples and their way of life.

Rafael Correa has stated that the country's first option is to maintain the crude oil in the subsoil. The national and international communities would be called on to help the Ecuadorian government implement this costly decision for the country. The government hopes to recover 50 per cent of the revenues it would obtain by extracting the oil. The procedure involves the issuing of government bonds for the crude oil that will remain in situ, with the double commitment of never extracting this oil and of protecting Yasuni National Park. It is important to keep in mind that if Ecuador succeeds in receiving the hoped for amount – estimated at US$350 million annually – it would only be for a period of ten years beginning after the sixth year, since production and potential revenues would progressively decline at the end of that period.

A more promising alternative would be a strategy to provide the government with the 50 per cent of resources in such a way as to provide a consistent income for an indefinite period of time. These resources would be channelled towards activities that help to free the country from its dependency on exports and imports and to consolidate food sovereignty. The proposal is framed within the national and international contexts based on the following considerations:

1. halt climate change
2. stop destruction of biodiversity
3. protect the Huaorani people
4. economic transformation of the country.

The very notion of an 'ecological debt' is also a non-reformist reform, because although it asserts the calculation of the monetary value of nature, payment on such an obligation would revise such a range of power relationships that massive structural change would inevitably follow. Such linkages between environmental stewardship and social justice provide the only sure way to generate political principles that can inform lasting climate mitigation.

We need to review the proposals in between and ask: will principles of non-reformist reformism be adopted by those advocating carbon taxes and per-capita emissions rights? Two crucial questions emerge which will help determine whether the reforms proposed by carbon tax and per-capita emissions rights advocates do more harm than good. The first is whether the kinds of reforms proposed – which entail putting a price on carbon and exposing that price (and all manner of related negotiations) to corporate-dominated national and global-scale 'governance' initiatives – can be assured of both genuinely addressing the climate crisis and also redistributing energy and economic resources from rich to poor. The devil is in the details in relation to both a carbon tax and per-capita emissions rights, yet as noted, the presumptions entailed in taxation (which often has a maldistributive impact, as shown in the British Columbia gas tax) and allocations of property rights will make a constructive outcome unlikely.

We are left asking, as a result, whether non-reformist reform opportunities might emerge so that a carbon tax can redistribute resources to both renewable energy investments and to low-income people who, through no fault of their own, are most vulnerable to higher energy prices. Could a per-capita rights mechanism be designed and adopted that moves forward the agenda of the environmental and social justice movements without falling victim to market distortions? These are not impossible outcomes, but given prevailing power relations, they are quite unlikely.

The second question is whether pursuing these sorts of reforms will contribute to the expansion and empowerment of the environmental justice movement. At the December 2007 Bali Conference of Parties, a movement emerged to unite 'green' and 'red' demands:

- reduced consumption;
- huge financial transfers from North to South based on historical responsibility and ecological debt for adaptation and mitigation costs paid for by redirecting military budgets, innovative taxes and debt cancellation;
- leaving fossil fuels in the ground and investing in appropriate energy efficiency and safe, clean and community-led renewable energy;
- rights-based resource conservation that enforces indigenous land rights and promotes peoples' sovereignty over energy, forests, land and water; and
- sustainable family farming and people's food sovereignty.

The alternative strategies proposed above do not rely entirely upon command-and-control, which requires national and ultimately global state power, which is not likely to be exercised by environmentally responsible political parties for many years (if not decades) notwithstanding encouraging signs from Ecuador. Instead, a new approach to command-and-control from below is being adopted, which takes forward community, labour and environmental strategies to maintain resources in the ground, especially fossil fuels and especially in cases where 'resource curse' economic power relations prevail. It is in such cases where activists have an unprecedented opportunity.

Because of the failure of elites to properly recognise and address climate change, and because their strategy of commodifying the commons through the CDM was already a serious threat to numerous local communities across the Third World, the Durban Group for Climate Justice produced a Declaration on Carbon Trading in October 2004, which rejected the claim that this strategy could halt the climate crisis. It insisted that the crisis has been caused more than anything else by the mining of fossil fuels and the release of their carbon to the oceans, air, soil and living things.

Conclusion: Direct action to protect the climate commons

It is here, finally, where the most crucial lesson of the climate debate lies: in confirming the grassroots, coalface and fence-line demand by civil society activists to leave the oil in the soil, the coal in the hole, the resources in the ground. This demand emanated in a systemic way at the Kyoto Protocol negotiations in 1997 from the group OilWatch when it was based in Quito, Ecuador, as heroic activists from Acción Ecológia took on struggles such as halting exploitation of the Yasuni oil.

Within a decade, in January 2007, at the World Social Forum in Nairobi, many other groups became aware of this movement thanks to eloquent activists from the Niger Delta, including the Port Harcourt NGO, Environmental Rights Action (ERA). The ERA visited Durban in March 2007 to expand the network with allies, such as the South Durban Community Environmental Alliance and the Pietermaritzburg NGO groundWork, and these groups committed in July 2008 to campaign against the proposed pipeline from Durban to Johannesburg, which would double petrol product flow.

But the legacy of resisting fossil -fuel abuse goes back much further and includes Alaskan and Californian environmentalists who halted drilling and even exploration. In Norway, the global justice group, Association for the Taxation of Financial Transactions for the Aid of Citizens (ATTAC) took up the same concerns in an October 2007 conference and began the hard work of persuading wealthy Norwegian oil fund managers that they should use the vast proceeds of their North Sea inheritance to repay Ecuadorans some of the ecological debt owed. In Australia, regular blockades of Newcastle coal transport (by rail and sea) are carried out by the activist group Rising Tide.

Canada is another Northern site where activists are hard at work to leave the oil in the soil. In a November 2007 conference in Edmonton, the Parkland Institute of the University of Alberta also addressed the need for no further development of tar sand deposits (which require a litre of oil to be burned for every three to be extracted, and which devastate local water, fisheries and air quality). Institute director, Gordon Laxer, laid out careful arguments for strict limits on the use of water and greenhouse gas emissions in tar sand extraction; realistic land reclamation plans (including a financial deposit large enough to cover full-cost reclamation up-front); no further subsidies for the production of dirty energy; provisions for energy security for Canadians (since so much of the tar sand extract is exported to the United States); and much higher economic rents on dirty energy to fund a clean energy industry (currently Alberta has a very low royalty rate). These kinds of provisions would strictly limit the extraction of fossil fuels and permit oil to leave the soil only under conditions in which much greater socio-ecological and economic benefit is retained by the broader society.

(I raised this issue in many sites in 2006–08, enthusiastically commenting on the moral, political, economic and ecological merits of leaving the oil in the soil. Unfortunately, in addition to confessing profound humility about the excessive fossil fuel burned by aeroplanes that have taken me on this quest, I must report on the only site where the message dropped like a lead balloon: Venezuela. At a July 2007 environmental seminar at the vibrant Centro Internacionale Miranda in Caracas, joined by the brilliant Mexican ecological economist David Barkin, our attempts failed to generate debate on whether petro-socialism might become a contradiction in terms.)

There are many other examples where courageous communities and environmentalists have lobbied successfully to keep non-renewable resources (not only fossil fuels) in the ground, for the sake of the environment, community stability, disincentivising political corruption and workforce health and safety. The highest-stake cases in South Africa at present may be the Limpopo Province platinum fields and Wild Coast titanium finds, where communities are resisting foreign companies. The extraction of these resources is incredibly costly in terms of local land use, water extraction, energy consumption and political corruption, and requires constant surveillance and community solidarity. Visiting the Wild Coast in August 2008, Minerals and Energy Affairs minister, Buyelwa Sonjica, illustrated how much was at stake, when she accused the Xolobeni community's lawyer of interfering with the Australian firm Mineral Resource Commodities' plan for titanium extraction:

> There is a man called Richard Spoor who has divided the community. He is a white person. Today Spoor says he is fighting for people's rights, but where was he when Joe Slovo (former South African Communist Party leader) was fighting for people's rights and was imprisoned for that? I ask myself: 'How much does he get for dividing our community? What is his agenda for not wanting progress in our community?' (cited in Kockott 2008)

This kind of statement is not an aberration, but reflects the South African government's willingness to buy into the North's agenda for the South's continued subordination. It is an integral part of a system – named 'global apartheid' by President Thabo Mbeki – that must be fully dismantled. What role are South African politicians and technocrats playing? Is it similar to that of the elite collaborators of the apartheid-era Bantustans?

In that case and in the promotion of carbon trade, the key point is self-interest. In September 2007, environment minister Marthinus van Schalkwyk told the International Emissions Trading Association (IETA) forum in Washington, DC, 'An all-encompassing global carbon market regime which includes all developed countries is the first and ultimate aim' (2007b). His government's 'National Climate Change Response Strategy' of 2004 reveals the basis for the bizarre prioritisation of raising more funds for CDM projects: 'It should be understood up-front that CDM primarily presents a range of commercial opportunities, both big and small. This could be a very important source of foreign direct investment, thus it is essential that the Department of Trade and Industry participate fully in the process' (see Appendix 1 in this volume).

This is the same government – led by Eskom and the Department of Trade and Industry – that has disconnected an estimated ten million low-income South Africans from electricity due to their inability to pay, while committing billions of rand of subsidies to yet another energy-guzzling aluminium smelter, at Coega in the Mandela Metropole (formerly Port Elizabeth). And while making these commitments, the leading politicians watch while Eskom regularly runs out of power generation capacity, leaving millions of people to suffer load-shedding brown-outs. The reason for the load-shedding, the government confesses now, is that international investors did not want to add private investment into the existing electricity grid – since wide-scale privatisation of power was anticipated during the late 1990s – at a time the price of electricity was the world's lowest. One crony-capitalist strategy negated another, it seems and while the two corporate elephants (privatisation and cheap power) tussled, people at grass roots were flattened.

Having failed to attract private investors (the US firm AES signed up in 2007 to build future eThekwini and Mandela Metropole power plants but soon retreated from commitments), the South African government could well have used the opportunity to change the electricity pricing strategy in a manner that would both address climate change and ensure a sufficient lifetime supply of free electricity for all South Africans. It did neither, but instead sought to amplify the problems of market-oriented energy by delving into the carbon market.

To propose commercial opportunities associated with carbon trading and, simultaneously, the intensification of South Africa's world-record carbon dioxide (CO_2) emissions, does have a certain logic. It is the logic of an immature, greedy society led by calculating, corrupt politicians and neoliberal technocrats – not a society that we can be proud to belong to.

The Durban Declaration on Carbon Trading rejected the claim that this strategy will halt the climate crisis. It insisted that the crisis has been caused more than anything else by the mining of fossil fuels and the release of their carbon to the oceans, air, soil and living things. The Declaration suggested that people need to be made more aware of carbon trading threat and to actively intervene against it. By August 2005, inspiring citizen activism in Durban's Clare Estate community forced the eThekwini municipality to withdraw an application to the World Bank for carbon trading finance to include methane extraction from the vast Bisasar Road landfill (instead, the application was for two relatively small eThekwini dumps).

But the heroic battle against Bisasar's CDM status was merely defensive and the loss of Sajida Khan to cancer in July 2007 was a great blow to the struggle there. Community residents have a proactive agenda, to urgently ensure the safe and environmentally sound extraction of methane from the Bisasar Road landfill, even if that means slightly higher rubbish removal bills for those in Durban who are thoughtlessly filling its landfills, without recycling their waste. Khan's brother Rafiq will pick up Sajida's banner.

Clare Estate's apartheid-era dump should now finally be closed, a decade after originally promised. Simultaneously, good jobs and bursaries should be given to the dump's neighbours, especially in the Kennedy Road community, as partial compensation for their long suffering. Their fight for housing and decent services has been equally heroic; the current handful of toilets and standpoints for six thousand people should shame the eThekwini municipal officials, whose reprehensible response was to mislead residents into believing dozens of jobs would materialise through World Bank CDM funding.

A commitment is also needed to a zero-waste philosophy and policies by eThekwini and all other municipalities in South Africa. In Bellville, Western Cape, solidarity is needed for the many residents who are also victims of apartheid-dumping and who may also be victimised by the Bellville landfill's status as a CDM project.

Allies are needed in South African, African and international civil society. In October 2004, only cutting-edge environmental activists and experts understood the dangers of carbon trading. Others – including many well-meaning climate activists – argued that the dangers are not intrinsic in trading, just in the rotting low hanging fruits that represent the first and easiest projects to fund, at the cheapest carbon price.

Since then, however, numerous voices have been raised against carbon colonialism. These voices oppose the notion that, through carbon trading, Northern polluters can continue their fossil-fuel addiction, drawing down the global atmospheric commons in the process.

Rather than foisting destructive schemes like the toxic Bisasar Road dump on the South, the North owes a vast ecological debt. For playing the role of carbon sink alone, political ecologist Joan Martinez-Alier and United Nations (UN) climate change commissioner Jyoti Parikh calculate that an annual subsidy of US$75 billion is provided from South to North.

In October 2004, the Durban Group noted that the internal weaknesses and contradictions of carbon trading are likely to make global warming worse, rather than mitigate it. We are ever more convinced of this in South Africa, partly because in mid-2005, a leading official of state-owned Sasol publicly conceded that his own ambitious carbon-trading project is merely a gimmick, without technical merit because he cannot prove what is termed 'additionality'. The crony character of the CDM verification system may allow this travesty to pass into the market, unless our critique is amplified.

In October 2004, we worried that giving carbon a price through the emissions market would not prove to be any more effective, democratic, or conducive to human welfare, than giving genes, forests, biodiversity or clean rivers a price. Over the past years, the South African government's own climate change strategy has been increasingly oriented itself to the commercial opportunities associated with carbon.

Worse, as South Africa often does in Africa, the government's agenda appears to be legitimisation of climate neoliberalism. As Van Schalkwyk commented in September 2006 in preparation for the Nairobi COP to the Kyoto Protocol, 'To build faith in the carbon market and to ensure that everyone shares in its benefits, we must address the obstacles that African countries face'.[1] This was the second of his three main priorities for 'action' in Nairobi, along with adaptation and stronger targets for emissions reductions – while South Africa continues its own irresponsible trajectory of energy-intensive, fossil-fuelled corporate subsidisation.

In the December 2007 COP meeting, Van Schalkwyk acquired a certain kind of Third World and African leadership, as a G77 negotiator aiming to pull the United States into the market-oriented climate regime, a strategy that proved only marginally successful, given that the price was the evacuation of any emissions target and accountability mechanism in the official declaration.

That reputation carried over into the climate scenarios noted in the Introduction, in which large emissions cuts are announced, yet virtually no mention is made of the extraordinary emissions increases that are planned in coming years through the vast new coal-fired electricity generation plants under construction. In this context, Van Schalkwyk's officials, including deputy director-general, Joanne Yawitch, have the confidence to enter power politics, on behalf of other major emitters who want to resist European pressure to cut greenhouse gases. In Accra in September 2008, Yawitch took up from where Van Schalkwyk left off, adding the crucial point that even CDM investments – by no means obviously a failure – were insufficient to bribe South Africa into endorsing genuine emissions cuts:

> Indian and Chinese negotiators accuse rich countries, notably those in the European Union, of attempting to divide the developing countries – known as the G77+China – and nudging some to accept legally binding limits on carbon emissions, despite the Kyoto Protocol not compelling them to do so. According to Indian and Chinese negotiators South Africa – which, after Nigeria, has the highest carbon emissions on the continent – is being promised more clean development projects if it strays from the consolidated stance of the G77. South Africa has the highest number of clean development mechanism projects in Africa. Yawitch says such attempts to isolate South Africa are fruitless because these processes have not worked for Africa. 'There are big problems with the CDMs. There are capacity issues and often the countries cannot afford the transaction costs [of researching and submitting proposals].' She says that South Africa has very few CDMs in comparison to

> places such as Brazil and China. The G77+China as a bloc has committed to not making any individual commitments on targets for carbon emissions, says Yawitch. 'There might be attempts to divide us, because there are divisions within the different groups in the G77. But we are clear that none of us are talking about binding targets.' (Rossouw 2008)

The bottom line is that the South African ruling elite remains addicted to fossil fuels and is happy to ally with China and India to maintain its addiction. In addition to a massive – and partly privatised – increase in cheap coal-fired electricity generation for the sake mainly of large corporations, the South African government also exhibits a range of other irresponsible energy policies, on which Van Schalkwyk is silent (or worse, laudatory), including inadequate subsidies and research and development commitments to renewable energy; a renewed focus on nuclear energy using the specious, incorrect argument that it is safer, cheaper and cleaner than coal and a turn to potential hydroelectricity projects, which even the South African-based World Commission on Dams condemned as often contributing more to global warming than coal-generated electricity through methane emissions from plant decay.

One of the most eloquent climate analysts is George Monbiot, so it was revealing that in December 2007, instead of going to Bali, he stayed home in Britain and caused some trouble, reporting back in his *Guardian* column:

> Ladies and gentlemen, I have the answer! Incredible as it might seem, I have stumbled across the single technology which will save us from runaway climate change! From the goodness of my heart I offer it to you for free. No patents, no small print, no hidden clauses. Already this technology, a radical new kind of carbon capture and storage, is causing a stir among scientists. It is cheap, it is efficient and it can be deployed straight away. It is called . . . leaving fossil fuels in the ground.
>
> On a filthy day last week, as governments gathered in Bali to prevaricate about climate change, a group of us tried to put this policy into effect. We swarmed into the opencast coal mine being dug at Ffos-y-fran in South Wales and occupied the excavators, shutting down the works for the day. We were motivated by a fact which the wise heads in Bali have somehow missed: if fossil fuels are extracted, they will be used . . . The coal extracted from Ffos-y-fran alone will produce 29.5 million tonnes of carbon dioxide equivalent, according to the latest figures from the Intergovernmental Panel on Climate Change, to the sustainable emissions of 55 million people for one year . . .

> Before oil peaks, demand is likely to outstrip supply and the price will soar. The result is that the oil firms will have an even greater incentive to extract the stuff.
>
> Already, encouraged by recent prices, the pollutocrats are pouring billions into unconventional oil. Last week BP announced a massive investment in Canadian tar sands. Oil produced from tar sands creates even more carbon emissions than the extraction of petroleum. There's enough tar and kerogen in North America to cook the planet several times over.
>
> If that runs out they switch to coal, of which there is hundreds of years' supply. Sasol, the South African company founded during the apartheid period (when supplies of oil were blocked) to turn coal into liquid transport fuel, is conducting feasibility studies for new plants in India, China and the US. Neither geology nor market forces is going to save us from climate change.
>
> When you review the plans for fossil fuel extraction, the horrible truth dawns that every carbon-cutting programme on earth is a con. Without supply-side policies, runaway climate change is inevitable, however hard we try to cut demand. (Monbiot 2007a)

Real solutions to the climate crisis are needed and with its world-leading CO_2 emissions, South Africa must be at the cutting edge of progressive climate activism, not a leading partner in the privatisation of the atmosphere. This, in turn, will require the resolution of another vast challenge: the lack of synthesis between the three major citizens' networks that have challenged government policy and corporate practices: environmentalists, community groups and trade unions. More work is required to identify the numerous contradictions within both South African and global energy sector policies and practices and to help to synthesise the emerging critiques and modes of resistance within progressive civil society. Only from this process of praxis can durable knowledge be generated about how to solve the climate and energy crises in a just way.

Finally, consider the message that Al Gore provided privately to a *New York Times* columnist in August 2007: 'I can't understand why there aren't rings of young people blocking bulldozers and preventing them from constructing coal-fired power plants.' (Kristoff 2007) The suggestion should be heard in South Africa.

Note

1. See http://www.info.gov.za/speeches/2006/06091112451001.htm.

Appendix 1

South Africa's Clean Development Mechanism Policy

National Climate Change Response Strategy, September 2004

Department of Environmental Affairs and Tourisn

South Africa, as a non-Annex 1 country, is not required to reduce its emissions of greenhouse gases. However, the South African economy is highly dependent on fossil fuels and the country can be judged to be a significant emitter due to the relatively high values that can be derived for emissions intensity and emissions per capita. Such calculations put South Africa as one the world's top fifteen most energy intensive economies, with a significant contribution to greenhouse emissions at a continental level.

There could be benefits to be derived from adopting a future strategy that is designed to move the economy towards a cleaner development path. This will further require development of a strategy to access investment through the Clean Development Mechanism (CDM) of the Kyoto Protocol, technology transfer and donor funding opportunities. However, even given this scenario, emissions can still be expected to increase with economic development, albeit at a smaller pace than would have happened without intervention . . .

Government urgently needs to establish procedures for the registration, coordination and reporting on projects to be undertaken through the CDM. Detailed discussions have been held between high-level delegations from DEAT, DTI and DME [Departments of Environmental Affairs and Tourism, Trade and Industry and Minerals and Energy]. The following mechanisms are proposed, are being developed or have already been instituted:

a) A CDM secretariat is being set up within DME and it is envisaged that the Director General of DME will act, for legal purposes, as the Designated National Authority (DNA) in terms of the Kyoto Protocol, in which capacity he will have full signing authority and the associated accountability.

b) The DNA will be advised by a steering committee, chaired jointly by DME, DEAT and the DTI. It is, however, essential that other departments (for example the Department of Foreign Affairs) be permanently represented on the committee, as should other stakeholders, including civil society.
c) The CDM secretariat will introduce proposals to the steering committee who will make recommendations to the DNA. The DNA will issue letters of approval.
d) DTI would provide guidance on possible trade and investment implications of projects and will assist in the marketing of potential CDM projects in South Africa.
e) DTI will be instrumental in ensuring that, where possible, the CDM is used to support national trade and investment measures.
f) The CDM secretariat would provide a single point of entry for all information pertaining to the CDM, and would be able to advise on all aspects of the necessary South African and international processes and requirements.
g) The secretariat would be responsible for the registration of all projects, but not for actual project management, which would be the responsibility of the project developers.
h) The secretariat would serve as a focal point to the CDM Executive Board, as set up under the Kyoto Protocol, and deal with correspondence from this Board.
i) The secretariat would also provide input into the negotiating process on the CDM, through the NCCC.
j) The arrangements could be considered as interim with the possibility of them being reviewed in light of performance achieved, status of the Kyoto Protocol negotiations and the future scale of the CDM market.

It should be understood up-front that CDM primarily presents a range of commercial opportunities, both big and small. This could be a very important source of foreign direct investment, thus it is essential that the DTI participate fully in the process. Contracting organisations from the recipient country can range from large private corporations, parastatals and the smaller commercial operations of academic institutes and consultancies aligned with NGOs [non-governmental organisations]. The actual range of potential projects is very large and cannot be covered in detail here. However, as just a few examples, they could encompass fuel switching from coal to gas, clean coal technologies, energy efficient housing, the use of renewable energy resources or the production of electricity from landfill gas, as well as numerous other applications. The identification of suitable projects could be assisted by the results of the technology needs analysis referred to elsewhere in this document.

The overall governance and coordination of CDM is through the CDM Executive Board established under the Kyoto Protocol. The responsibility for constituting and appointing the Executive Board lies with the UNFCCC [United Nations Framework Convention on Climate Change] conference of parties/meeting of parties structures. There are mechanisms to ensure equitable regional representation and a balance between developed and developing nation representation. The Executive Board is mandated with the administration of an adaptation fund to oversee allocations to adaptation projects, specifically for the poorest and most vulnerable nations, with the prioritisation of funding in accordance with criteria established from the vulnerability assessments submitted to the UNFCCC conference of parties.

All information should be entered into a project information management system. However, the Secretariat could keep all proprietary information confidential at all stages. The detailed evaluation of greenhouse gas reductions needs to be done according to standard methodologies as laid down by the Executive Board and through the Designated Operational Entities (DOEs) mandated by the Board. On applying for pre-approval, a 2-month turn around time, or shorter, should be guaranteed. The project should be evaluated for economic benefits, social benefits, and technological feasibility. The public will be consulted on the sustainable development criteria, which can be unique for South Africa. The process for the application of these criteria will be specified. The primary role of the CDM process is to assess projects against these sustainable development criteria, but those responsible will require the necessary information in order for them to do this. Technical feasibility could be evaluated through using specific members of an expert panel who have been chosen for their technical competence and willingness to respond rapidly. It is doubtful whether adequate capacity in this area would normally reside within the DNA and/or the steering committee or secretariat. However, the composition and role of this panel will need to be clearly defined as to the required level of their assessment and their terms of reference should be limited to that of acting in an advisory capacity only.

The expert panel would not be required to sit formally and review projects. Projects could be referred to the appropriate experts by email. In cases that require an Environmental Impact Assessment (EIA), then a process of public participation will, in any event, need to be conducted and various stakeholders consulted. This type of process should not be duplicated, as it will inevitably result in the process becoming even lengthier. It should be noted that the risk with regards to obtaining approval of EIA's is borne by the project developers and the EIA could be carried out in advance of the CDM approval process should the project developer wish to do so. In addition the initiation or carrying out of an EIA should not be considered as invalidating the proposed project on the grounds that it represents 'business as usual'. The application for full approval should contain complete project specifications and a detailed account of the proposal for verifying the emissions reductions. The CDM Executive Board in Washington is likely to make approval conditional upon continued achievement of requirements. This process should not take longer than 14 weeks from start to finish, preferably much less, excluding the time taken to process the EIA, where necessary.

The allocation of certified emission reductions has not as yet been finalised. However, it is widely thought that ownership would essentially remain with the project developers to give the incentive to carry out CDM projects, with governments retaining overall custodianship of the national interests. It is expected that the CDM Executive Board would maintain a CDM registry and that South Africa, as the host party, as well as the project participants would have registry accounts into which certified emission reductions would be transferred directly by the CDM Executive Board.

Appendix 2

The Durban Declaration on Carbon Trading

The Durban Group for Climate Justice

As representatives of people's movements and independent organisations, we reject the claim that carbon trading will halt the climate crisis. This crisis has been caused more than anything else by the mining of fossil fuels and the release of their carbon to the oceans, air, soil and living things. This excessive burning of fossil fuels is now jeopardising Earth's ability to maintain a liveable climate.

Governments, export credit agencies, corporations and international financial institutions continue to support and finance fossil fuel exploration, extraction and other activities that worsen global warming, such as forest degradation and destruction on a massive scale, while dedicating only token sums to renewable energy. It is particularly disturbing that the World Bank has recently defied the recommendation of its own Extractive Industries Review which calls for the phasing out of World Bank financing for coal, oil and gas extraction.

We denounce the further delays in ending fossil fuel extraction that are being caused by corporate, government and United Nations' attempts to construct a 'carbon market', including a market trading in 'carbon sinks'.

History has seen attempts to commodify land, food, labour, forests, water, genes and ideas. Carbon trading follows in the footsteps of this history and turns the earth's carbon-cycling capacity into property to be bought or sold in a global market. Through this process of creating a new commodity – carbon – the Earth's ability and capacity to support a climate conducive to life and human societies is now passing into the same corporate hands that are destroying the climate.

People around the world need to be made aware of this commodification and privatisation and actively intervene to ensure the protection of the Earth's climate. Carbon trading will not contribute to achieving this protection of the Earth's climate. It is a false solution which entrenches and magnifies social inequalities in many ways:

- The carbon market creates transferable rights to dump carbon in the air, oceans, soil and vegetation far in excess of the capacity of these systems to hold it. Billions of dollars worth of these rights are to be awarded free of charge to the biggest corporate emitters of greenhouse gases in the electric power, iron and steel, cement, pulp and paper, and other sectors in industrialised nations who have caused the climate crisis and already exploit these systems the most. Costs of future reductions in fossil fuel use are likely to fall disproportionately on the public sector, communities, indigenous peoples and individual taxpayers.
- The Kyoto Protocol's Clean Development Mechanism (CDM), as well as many private sector trading schemes, encourage industrialised countries and their corporations to finance or create cheap carbon dumps such as large-scale tree plantations in the South as a lucrative alternative to reducing emissions in the North. Other CDM projects, such as hydrochlorofluorocarbons (HCFC) reduction schemes, focus on end-of-pipe technologies and thus do nothing to reduce the impact of fossil fuel industries' impacts on local communities. In addition, these projects dwarf the tiny volume of renewable energy projects which constitute the CDM's sustainable development window-dressing.
- Impacts from fossil-fuel industries and other greenhouse-gas producing industries such as displacement, pollution, or climate change, are already disproportionately felt by small island states, coastal peoples, indigenous peoples, local communities, fisherfolk, women, youth, poor people, elderly and marginalised communities. CDM projects intensify these impacts in several ways. First, they sanction continued exploration for, and extraction, refining and burning of fossil fuels. Second, by providing finance for private sector projects such as industrial tree plantations, they appropriate land, water and air already supporting the lives and livelihoods of local communities for new carbon dumps for Northern industries.
- The refusal to phase out the use of coal, oil and gas, which is further entrenched by carbon trading, is also causing more and more military conflicts around the world, magnifying social and environmental injustice. This in turn diverts vast resources to military budgets which could otherwise be utilised to support economies based on renewable energies and energy efficiency.
- In addition to these injustices, the internal weaknesses and contradictions of carbon trading are in fact likely to make global warming worse rather than 'mitigate' it. CDM projects, for instance, cannot be verified to be 'neutralising' any given quantity of fossil fuel extraction and burning. Their claim to be able to do so is increasingly dangerous because it creates the illusion that consumption and production patterns, particularly in the North, can be maintained without harming the climate.
- In addition, because of the verification problem, as well as a lack of credible regulation, no one in the CDM market is likely to be sure what they are buying. Without a viable

commodity to trade, the CDM market and similar private sector trading schemes are a total waste of time when the world has a critical climate crisis to address.

- In an absurd contradiction the World Bank facilitates these false, market-based approaches to climate change through its Prototype Carbon Fund, the BioCarbon Fund and the Community Development Carbon Fund at the same time it is promoting, on a far greater scale, the continued exploration for, and extraction and burning of fossil fuels – many of which are to ensure increased emissions of the North.

In conclusion, 'giving carbon a price' will not prove to be any more effective, democratic, or conducive to human welfare, than giving genes, forests, biodiversity or clean rivers a price.

We reaffirm that drastic reductions in emissions from fossil fuel use are a prerequisite if we are to avert the climate crisis. We affirm our responsibility to coming generations to seek real solutions that are viable and truly sustainable and that do not sacrifice marginalised communities. We therefore commit ourselves to help build a global grassroots movement for climate justice, mobilise communities around the world and pledge our solidarity with people opposing carbon trading on the ground.

10 October 2004
Glenmore Centre, Durban, South Africa

Durban Meeting Signatories

Carbon Trade Watch; Indigenous Environmental Network; Climate & Development Initiatives, Uganda; Coecoceiba-Amigos de la Tierra, Costa Rica; CORE (Centre for Organisation Research & Education), Manipur, India; Delhi Forum, India; Earthlife Africa (ELA) eThekwini Branch, South Africa; FERN, EU; FASE-ES/Green Desert Network Brazil; Global Justice Ecology Project, USA; groundWork, South Africa; National Forum of Forest People And Forest Workers (NFFPFW), India; Patrick Bond, Professor, University of KwaZulu-Natal School of Development Studies, South Africa; O le Siosiomaga Society, Samoa; South Durban Community Alliance (SDCEA), South Africa; Sustainable Energy & Economy Network, USA; The Corner House, UK; Timberwatch Coalition, South Africa; World Rainforest Movement, Uruguay.

Supporting Organisational Signatories

50 Years is Enough: U.S. Network for Global Economic Justice, USA; Aficafiles, Canada; Africa Groups of Sweden, Sweden; Alianza Verde, Honduras; Ambiente y Sociedad, Argentina; Angikar Bangladesh Foundation, Bangladesh; Anisa Colombia, Colombia; Asociacion Alternativa Ambiental, Spain; Asociación Amigos Reserva Yaguaroundi, Argentina; Asociación de Guardaparques Argentinos, Argentina; Asociación Ecologista

Piuke, Argentina; Asociación para la Defensa del Medio Ambiente del Noreste Santafesino, Argentina; Asociación San Francisco de Asís, Argentina; Association France Amerique Latine, France; Associación Lihue San Carlos de Barloche/Rio Negro, Argentina; Association pour un contrat mondial de l'eau, Comité de Seine Saint Denis, France; Associação Caeté – Cultura e Natureza, Brazil; Athlone Park Residents Association, South Africa; Austerville Clinic Committee, South Africa; Australian Greens, Australia; Auckland Rising Tide, New Zealand; BanglaPraxis, Bangladesh; Benjamin E. Mays Center, USA; Bluff Ridge Conservancy (BRC), South Africa; BOA, Venezuela; Boulder Environmental Activists Resource, Rocky Mountain; Peace and Justice Center, USA; The Bread of Life Development Foundation, Nigeria; CENSAT – Friends of the Earth Colombia, Colombia; Center for Economic Justice, USA; Centre for Environmental Justice, Sri Lanka; Center for Environmental Law and Community Rights Inc.; Friends of the Earth (PNG), Papua New Guinea; Center for Urban Transformation, USA; Centro de Derecho Ambientaly Promoción para el Desarrollo (CEDAPRODE), Nicaragua; Centro de Investigacion Cientifica de Yucatan A.C., Mexico; Committee in Solidarity with the People of El Salvador, USA; Christ the King Church Group, South Africa; Clairwood Ratepayers Association (CRA), South Africa; Cold Mountain, Cold Rivers, USA; Colectivo de Proyectos Alternativos de México (COPAL), Mexico; Colectivo MadreSelva, Guatemala; Comité de Análisis 'Ana Silvia Olán' de Sonsonate – CANASO, El Salvador; Committee in Solidarity with the People of El Salvador, USA; Community Health Cell, Bangalore, India; Corporate Europe Observatory (CEO), Netherlands; C.P.E.M. Nº29-Ciencias Ambientales, Argentina; Del Consejo de Organisaciones de Médicos y Parteras Indígenas Tradicionales de Chiapas, Mexico; Enda América Latina, Colombia; ECOGRAIN, Spain; Ecoisla, Puerto Rica; EarthLink e.V. – The People & Nature Network, Germany; Ecological Society of the Philippines, Philippines; Ecologistas en Acción, Spain; Ecoportal.net, Argentina; ECOTERRA International; El Centro de Ecología y Excursionismo de la Universidad de Carabobo, Venezuela; Els Verds – Alternativa Verda, Spain; Environment Desk of Images Asia, Thailand; FASE Gurupá, Brazil; Forest Peoples Programme, UK; Foundation for Grassroots Initiatives in Africa, Ghana; Friends of the Earth International; Friends of the Earth Australia, Australia; Friends of the Siberian Forests, Russia; FSC-Brazil, Brazil; Fundación Argentina de Etoecología (FAE), Argentina; Fundación Los de Tilquiza, proyecto AGUAVERDE, Argentina; Groupe d'Etudes et de Recherche sur les Energies Renouvelables et l'Environnement (GERERE), Morocco; Gruppo di Volontariato Civile (GVC – Italia), Italy; oficina de Nicaragua, Nicaragua; House of Worship, South Africa; Indigenous Peoples' Biodiversity Network, Peru; InfoNature, Portugal; Infringement Festival, Canada; Iniciativa ArcoIris de Ecologia y Sociedad, Argentina; Iniciativa Radial, Argentina; Institute for Social Ecology Biotechnology Project, USA; Instituto Ecoar para Cidadania, Brazil; Instituto Igaré, Brazil; International Fund for Animal Welfare (IFAW), Belgium; International Indian Treaty

Council; Isipingo Environmental Committee (IEC), South Africa; Isipingo Ratepayers Association, South Africa; Jeunesse Horizon, Camerun; JKPP/Indonesian Community Mapping Network, Indonesia; Joint Action Committee of Isipingo (JACI), South Africa; KVW Translations, Spain; LOKOJ, Bangladesh; London Rising Tide, UK; Malvarrosamedia, Spain; Mangrove Action Project (MAP), USA; Mano Verde, Colombia; Mercy International Justice Network, Kenya; Merebank Clinic Committee (MCC), South Africa; Movimiento por la Paz y el Ambiente, Argentina; Movimento por los Derechos y la Consulta Ciudadana, Chile; Nicaragua Center for Community Action, USA; Nicaragua Network (US), USA; Nicaragua-US Friendship Office, USA; NOAH – Friends of the Earth Denmark, Denmark; Núcleo Amigos da Terra, Brazil; Ogoni Rescue Patriotic Fund, Nigeria; OilWatch International, Ecuador; OilWatch Africa, Nigeria; Organisacion Fraternal Negra Honduirena, Honduras; Parque Provincial Ernesto Tornquist, Argentina; Pacific Indigenous Peoples Environment Coalition (PIPEC), Aotearoa/New Zealand; Pesticides Action Network Latin America, Uruguay; Piedad Espinoza Trópico Verde, Guatemala; PovoAção, Brazil; Prideaux Consulting, USA; Projeto tudo Sobre Plantas – Jornal SOS Verde, Brazil; Public Citizen, USA; Rainforest Action Network, USA; Rainy River First Nations, Canada; Reclaim the Commons, USA; Red de Agricultura Orgánica de Misiones, Argentina; REDES – Amigos de la Tierra, Uruguay; Red Verde, Spain; Rettet den Regenwald, Germany; Rising Tide, UK; Sahabat Alam Malaysia/FOE – Malaysia, Malaysia; San Francisco Bay Area Jubilee Debt Cancellation Coalition, USA; Scottish Education and Action for Development, UK; S.G. Fiber, Pakistan; Silverglen Civic Association (SCA), South Africa; Sisters of the Holy Cross – Congregation Justice Committee, USA; Sobrevivencia, Friends of the Earth Paraguay, Paraguay; Sociedad Civil, Mexico; SOLJUSPAX, Philippines; Tebtebba Foundation, Philippines; The Sawmill River Watershed Alliance, USA; TRAPESE – Take Radical Action Through Popular Education and Sustainable Everything, UK/Spain; Treasure Beach Environmental Forum (TBEF), South Africa; Uganda Coalition for Sustainable Development, Uganda; Ujamaa Community Resource Trust (UCRT), Tanzania; UNICA, Nicaragua; Union Chrétienne pour l'Education et Développement des Déshérités (UCEDD), Burundi; Union Mexicana de Emprendedores Inios, A.C., Mexico; VALL DE CAN MASDEU, Spain; Wentworth Development Forum (WDF), South Africa; Western Nebraska Resources Council, USA; World Bank Boycott/Center for Economic Justice, USA; worldforests, UK; World Peace Prayer Society, USA.

Individual Signatories

Aarran Thomson, USA; Ángeles Leonardo, Argentina; Arlex González Herrera, Colombia; Beth Burrows, USA; Dr Bob de Laborde, South Africa; Brook Goldzwig, USA; Cesar Antonio Sanchez Asian, Peru; Christopher Keene, UK; Cláudia Sofia Pereira Henriques, Portugal; Claudio Capanema, Brazil; Daniel Tietzer, US; Dany Mahecha Rubio, the

Netherlands; Dora Fernandes, Portugal; Dulce Delgado, Portugal; Eduardo Rojas Hidalgo, Ecuador; Edwin S. Wilson, USA; Eileen Wittewaal, Canada; Elisa Marques, Portugal; Emmanuel Moutondo, Kenya; Fabry Saavedra, Bolivia; Federico Ivanissevich, Argentina; Florencia T. Cuesta, Argentina; Florian Salazar-Martin, France; Fernando Moran, Spain; Fernando Guzmán, Peru; Gar W. Lipow, USA; German A. Parra Bustamente, Colombia; Hannes Buckle, South Africa; Hansel Tietzer, USA; Helena Pinheiro, Brazil; Dr Hugh Sanborn, USA; Hylton Alcock, South Africa; Hsun-Yi Hsieh, Taiwan; Inês Vaz Rute da Conceição, Portugal; Irina Maya, Portugal; Dr J. Gabriel Lopez, USA; James Mabbitt, UK; Jane Hendley, USA; Janet Weyker, USA; Javier Lizarraga, Uruguay; Jeff Purcell, USA; Jelena Ilic, Serbia & Montenegro; Jenny Biem, Canada; Joana Gois, Portugal; Joao Forte, Portugal; John Brabant, USA; Jonathan Derouchie, Canada; Joris Leemans, Belgium; Josep Puig, Spain; Joseph Herman, USA; Judith Amanthis, UK; Judith Vélez, Isla Verde, Puerto Rico; Karen Roothaan, USA; Karlee Rockey, USA; Kiki Goldzwig, USA; Laura Carlsen, IRC; Leonardo Ornella, Argentina; Lina Hällström, Sweden; Lorna Salzman, USA; Luis E. Silvestre, Puerto Rico; Luis Edoardo Sonzini Meroi, Nicaragua; Ing. Mabel Vullioud, Argentina; Manuel Pereira, Portugal; Marcelo Bosi de Almeida, Brazil; Maria Benedetti, Cayey, Puerto Rico; Maria de Fátima Marques, Portugal; Maria Fernanda Pereira, Colombia; María Jesús Conde, Spain; Dr María Luisa Pfeiffer, Argentina; Martha L. Downs, USA; Dr Martin Mowforth, UK; Mary Galvin, South Africa; Matheus Ferreira Matos Lima, Brazil; Maurice Tsalefac, Professor, Université de Yaoundé, Camerun; Michaeline Falvey, USA; Miguel Parra Olave, Chile; Mike Ballard, Australia; Mike Berry, UK; Nick Gotts, Scotland; Norbert Suchanek, Germany; Nuno Miguel O.P. Matos Sequeira, Portugal; Oya Akin, North Cyprus; Pablo Alarcón-Cháires, Mexico; Patrícia Angelo Batista, Portugal; Patricia Raynor, USA; Paulo Cesar Scarim, Brazil; Pedro Ribeiro, Portugal; Peter Rachleff, Professor, Macalester College, USA; Peter Sills, USA; Dr Philip Gasper, USA; Prakash Deshmukh, India; Priscila Lins P.F. do Amaral, Brazil; Rafael Arturo Acuña Coaquira, Bolivia; Rafael Chumbimune Zanabria, Peru; Rafael Renteria, USA; Raj Patel, South Africa; Ray Hajat, Malawi; Robin Clanahan, South Africa; Roger de Andrade, France; Rogerio M. Mauricio, Brazil; Roxana Mastronardi, Argentina; Ruth Zenger, Canada; Rufino Vivar Miranda, Mexico; Sajida Khan, South Africa; Sandra C. Carrillo, USA; Sara Hayes, USA; Saul Landau, USA; Sheila Goldner, USA; Sister Aloysia Zellmann, South Africa; Steve Wheeler, UK; Tobias Schmitt, Germany; Tyrell Haberkorn, USA; Usman Majeed, Canada; Wak Kalola, Canada; Zoraida Crespo Feliciano, Puerto Rico.

To sign this declaration, please send an e-mail to info@fern.org or visit http://www.sinkswatch.org.

Appendix 3

Climate Justice Now!

Coalition Founding Statement, Bali, 14 December 2007

During the Conference of the Parties to the United Nations Framework Convention on Climate Change held in Bali, Indonesia, in December 2007, a number of social movements and groups agreed to establish a coalition called Climate Justice Now! in order to enhance exchange of information and cooperation among themselves and with other groups with the aim of intensifying actions to prevent and respond to climate change.

Members of the coalition include Carbon Trade Watch, Transnational Institute; Center for Environmental Concerns; Focus on the Global South; Freedom from Debt Coalition, Philippines; Friends of the Earth International; Gendercc – Women for Climate Justice; Global Forest Coalition; Global Justice Ecology Project; Indonesia Civil Society Organizations Forum on Climate Justice; International Forum on Globalisation; Kalikasan-Peoples Network for the Environment; La Vía Campesina; members of the Durban Group for Climate Justice; OilWatch; Pacific Indigenous Peoples Environment Coalition, Aotearoa/New Zealand; Sustainable Energy and Economy Network; The Indigenous Environmental Network; Third World Network and World Rainforest Movement.

On 14 December, the coalition issued the following statement:

> Peoples from social organisations and movements from across the globe brought the fight for social, ecological and gender justice into the negotiating rooms and onto the streets during the UN climate summit in Bali. Inside and outside the convention centre, activists demanded alternative policies and practices that protect livelihoods and the environment.
>
> In dozens of side events, reports, impromptu protests and press conferences, the false solutions to climate change – such as carbon offsetting, carbon trading for forests, agrofuels, trade liberalisation and privatisation pushed by governments, financial institutions and multinational corporations – have been exposed.

Affected communities, Indigenous Peoples, women and peasant farmers called for real solutions to the climate crisis, solutions which have failed to capture the attention of political leaders. These genuine solutions include:

- reduced consumption;
- huge financial transfers from North to South based on historical responsibility and ecological debt for adaptation and mitigation costs paid for by redirecting military budgets, innovative taxes and debt cancellation;
- leaving fossil fuels in the ground and investing in appropriate energy-efficiency and safe, clean and community-led renewable energy;
- rights based resource conservation that enforces indigenous land rights and promotes peoples' sovereignty over energy, forests, land and water; and
- sustainable family farming and peoples' food sovereignty.

Inside the negotiations, the rich industrialised countries have put unjustifiable pressure on Southern governments to commit to emissions reductions. At the same time, they have refused to live up to their own legal and moral obligations to radically cut emissions and support developing countries' efforts to reduce emissions and adapt to climate impacts. Once again, the majority world is being forced to pay for the excesses of the minority.

Compared to the outcomes of the official negotiations, the major success of Bali is the momentum that has been built towards creating a diverse, global movement for climate justice.

We will take our struggle forward not just in the talks, but on the ground and in the streets – Climate Justice Now!

Select Bibliography

Aberman, M. 2006. 'Earthlife Africa Statement'. Cape Town, 14 August.

Agarwal, A. and S. Narain. 1991. *Global Warming in an Unequal World: A Case of Environmental Colonialism*. New Delhi: Centre for Science and Environment.

Agence France Press. 2008. 'Progress Falters on Road Map to New Climate Deal'. Bonn, Germany, 13 June.

Akhurst, M. 2002. 'Presentation by Head of Climate Change at BP Amoco'. Amsterdam, 19 February.

Alexander, N. 2004. 'Triage of Low-income Countries? The Implications of the IFI's Debt Sustainability Proposal'. http://www.servicesforall.org/html/otherpubs/judge_juryscore card. pdf.

Athanasiou, A. 2005. 'Greens Divided over Dissing Feds: Would Keeping Criticism Quiet Help?'. *Now*, 1 December.

Bailey, S. 2005. 'Alcan Will Probably Build $2,5 bln Smelter, IDC Says'. *Bloomberg News*, 13 July.

Bank Information Center, Bretton Woods Project, Campagna per la Riforma della Banca Mondiale, CEE Bankwatch Network, Friends of the Earth-International, Institute for Policy Studies, International Rivers Network, Oil Change International and Urgewald. 2006. 'How the World Bank's Energy Framework Sells the Climate and Poor People Short'. http://www.seen.org/PDFs/Energy_Framework_CSO.pdf

Basel Agency for Sustainable Energy (BASE). n.d. 'Gold Standard Backgrounder'. http://www.cdm goldstandard.org.

Batchelor, P. and S. Willett. 1998. *Disarmament and Defence Industrial Adjustment in South Africa*. Oxford: Oxford University Press.

Beall, J., O. Crankshaw and S. Parnell. 2002. *Uniting a Divided City: Governance and Social Exclusion in Johannesburg*. London: Earthscan.

Becker, E. and D. Sanger. 2005. 'Opposition to Doubling Aid for Africa'. *GreenLeft Weekly*, 2 June.

Beder, S. 1997. *Global Spin*. Devon: Green Book Ltd.

Bell, T. 2006. 'Wrong Idea of Consensus Steers Transnet Machine'. *Business Report*, 10 March.

Black, D. 2004. 'Democracy, Development, Security and South Africa's "Arms Deal"'. In *Democratising foreign policy? Lessons from South Africa*, eds. P. Nel and J. van der Westhuizen. Lanham, MD: Lexington Books.

Bloomfield, S. 2007. 'Why Britain's G8 Carbon Offsetting Pledge Rings Hollow in Cape Town'. *Independent on Sunday*, 27 January.

Bond, P. 1999. 'Globalisation, Pharmaceutical Pricing and South African Health Policy: Managing Confrontation with US Firms and Politicians'. *International Journal of Health Services*, 29 (4).

———. 2002. *Unsustainable South Africa: Environment, Development and Social Protest*. London: Merlin Press and Pietermaritzburg: University of Natal Press.

———. 2007. 'Zuma, the Centre-Left and the Left-Left'. *Counterpunch*, 17 December.

British Broadcasting Corporation (BBC). 2005. 'Q & A on the Kyoto Protocol'. http://www.news.bbc.co.uk/1/hi/sci/tech/4269921.stm.

Butler, R. 2008. '55% of the Amazon May be Lost by 2030. But Carbon-For-Conservation Initiatives Could Slow Deforestation'. http://www.mongabay.com, 23 January.

Caffentzis, G. 2004. 'The Petroleum Commons: Local, Islamic and Global'. *The Progress Report*. http://www.progress.org/2004/water26.htm.

Callendar, G.S. 1938. 'The Artificial Production of Carbon Dioxide and its Influence on Temperature'. *Quarterly Journal of the Royal Meteorological Society*, 64.

Cameron, J. 2005. 'Presentation to IETA'. UN Conference, Montreal, 5 December.

Carbon Trade Watch. 2003. 'The Sky is not the Limit: The Emerging Market in Greenhouse Gases'. TNI Briefing Series 1. Amsterdam: Transnational Institute. http://www.tni.org/ctw.

———. 2005. 'Hoodwinked in the Hothouse: The G8, Climate Change and Free-Market Environmentalism'. http://www.carbontradewatch.org/pubs/.

———. 2007a. 'Agrofuels – Towards a Reality Check in Nine Key Areas'. http://www. carbontrade watch.org/pubs/.

———. 2007b. 'The Carbon Neutral Myth: Offset Indulgences for your Climate Sins'. http: //www. carbontradewatch.org/pubs/.

CBS News. 2003. 'Mandela Slams Bush on Iraq'. 30 January. http://www.cbsnews.com/stories/2003/01/30/iraq/main538607.shtml.

CDMWatch and SinksWatch. 2004. 'How Plantar Sinks the World Bank's Rhetoric: Tree Plantations and the World Bank's Sinks Agenda'. http://www.sinkswatch.org.

'CEF Opens London Office for Carbon Trading'. *Mail & Guardian*, 5 October 2007. http: // www.carbon. org.za/news/news_item.asp?iID=70.

Centre for Science and the Environment. 2000. 'Carbon Colonialism'. *Equity Watch*, 25 October. http://www.cseindia.org/html/cmp/climate/ew/art20001025_4.htm.

Charles Anderson Associates. 1994. 'National Electricity Policy Synthesis Study, Vol. 1'. Report submitted to the Department of Mineral and Energy Affairs, 12 August.

City of Cape Town. 2004. 'Integrated Waste Management Plan Draft Assessment Report'. http:// www.capetown.gov.za.

Clarke, J. 1991. *Back to Earth: South Africa's Environmental Challenges*. Johannesburg: Southern Book Publishers.

Climate Care. 2004. 'Annual Report'. http://www.co2.org/news/climate-cares-2004-annual-report.

CNN. 2004. 'Mandela Extends Conciliatory Hand to United States'. 24 May. http://www.cnn.com.

Cogen, J. 2005. 'Presentation to IETA'. UN Conference, Montreal, 5 December.

Corporate Europe Observatory (CEO). 2000. 'Greenhouse Market Mania: UN Climate Talks Corrupted by Corporate Psuedo-solutions'. http://www.corporateeurope.org/greenhouse/index.html.

Crawford-Browne, T. 2004. 'The Arms Deal Scandal Review'. *African Political Economy*, 31: 329–42.

Creamer Media Engineering News. 2005. 'Eskom Will Seek to Cancel Commodity-linked Tariff Deals'. 29 June.

Dales, J. 1968. *Pollution, Property and Prices: An Essay in Policy-making and Economics*. Toronto: University of Toronto Press.

Daly, H. 2007. *Ecological Economics and Sustainable Development*. Cheltenham: Edward Elgar.

Daniel, J. and J. Lutchman. 2005. 'South Africa in Africa: Scrambling for Energy'. Presentation to the South African Association of Political Studies Colloquium, University of KwaZulu-Natal, Pietermaritzburg, 22 September.

Davie, K. 2007. 'Meet Mr Carbon'. *Mail & Guardian*, 14 September.

Den Elzen, M. and A. de Moor. 2001. 'Evaluating the Bonn Agreement and Some Key Issues'. Bilthoven: The National Institute of Public Health (RIVM). http://www.rivm.nl/bibliotheek/rapporten/728001016.html.

Department of Environmental Affairs and Tourism. 2005. 'Climate Change Science in Africa'. *Bojanala* (special issue on the National Climate Change Conference, Pretoria). http://www.environment.gov.za/HotIssues/2005/climateChange/docs/bojanala%20ed2.pdf.

———. 2008. 'Government's Vision, Strategic Direction and Framework for Climate Policy'. Pretoria, 28 July.

Department of Finance. 2001. '2002 Estimates of National Expenditure: Vote 30, Minerals and Energy'.

Department of Minerals and Energy. 1995. 'South African Energy Policy Document'.

———. 1997. 'Re-appraisal of the National Electrification Programme and the Formulation of a National Electrification Strategy'. http://www.dme.gov.za/energy/RE-APPRAISAL. htm.

———. 1998. 'White Paper on the Energy Policy of the Republic of South Africa'.

———. 2005. 'South Africa's Designated National Authority'.

Department of Provincial and Local Government. 2002. 'Quarterly Monitoring of Municipal Finances and Related Activities: Summary of Questionnaires for Quarter Ended 31 December 2001'.

Det Norske Veritas (DNV). 2002. 'Validation of the Plantar Project, Report No. 2001–1263'. 12 June.

Drury, R., M. Belliveau, J. Kuhn and S. Bansal. 1999. 'Pollution Trading and Environmental Injustice: Los Angeles' Failed Experiment in Air Quality Policy'. Duke Environmental Law and Policy Forum.

'Earth Summit Held in Brazil; Climate, Species Pacts Signed; Targets Lacking on Aid, Controls'. *Facts on File World News Digest*, 18 June 1992.

Earthlife Africa. 2001a. 'Nuclear Energy Costs the Earth'. Johannesburg: Congress of South African Trade Unions (COSATU).

———. 2001b. 'COSATU Submission on the Eskom Conversion Bill'. Presented to Public Enterprises Portfolio Committee, 9 May.

———. 2001c. 'Other Energy-related Developments'. Unpublished report.

———. 2002. 'Information Pack for Activists Training in Energy Issues'.

Eberhard, A. n.d. 'The Political, Economic, Institutional, and Legal Dimensions of Power Sector Reform in South Africa'. Presentation: Graduate School of Business, University of Cape Town and National Electricity Regulator.

Elbe, S. 2003. 'Strategic implications of HIV/AIDS'. *Adelphi Paper 357*. International Institute for Strategic Studies, Oxford: Oxford University Press.

Environmental Data Services (ENDS). 2002a. 'Report 327'. April.

———. 2002b. 'BP's Credibility Gap over Carbon Emissions'. Environmental Data Services Report, March.

———. 2003. 'Report 337'. February.

Environmental Defence Fund and Natural Resources Defence Council. 1994. 'Power Failure', report.

Environmental Media Services (EMS). 2003. 'The World Bank's Investments in Climate-changing Fossil Fuels'. 16 October.

Eraker, H. 2000. 'CO_2lonialism: Norwegian Tree Plantations, Carbon Credits and Land Use Conflicts in Uganda'. *NorWatch*, April.

Erion, G. 2005. 'Low Hanging Fruit Rots First'. In *Trouble in the Air*, eds. P. Bond and R. Dada. Durban: Centre for Civil Society, University of KwaZulu-Natal.

Fabricius, P. 2005. 'PetroSA to Send Technicians to Explore Oil Possibilities in the Sudan'. *The Star*, 5 January.

Fenhann, J. 2006. 'CDM Project Pipeline'. UNEP Risø Centre.

Fine, B. and Z. Rustomjee. 1996. *The Political Economy of South Africa: From Minerals-energy Complex to Industrialisation*. London: Christopher Hirst and Johannesburg: University of Witwatersrand Press.

Forest Stewardship Council (FSC). 2002. 'Evaluation Report of V&M Florestal Ltda. and Plantar S.A. Reflorestamentos, Both Certified by FSC'. Brazil, November. http://www. wrm.org.uy/countries/Brazil/fsc.html.

Fuggle, R. 2006. 'We are Still Indifferent about the State of our Environment'. *Cape Times*, 6 December.

Fyfe, G. 2004. 'Gas – the African Way'. *Global Energy Review*, June: 46.

Gary, I. and N. Reisch. 2005. *Chad's Oil: Miracle or Mirage?* Washington, DC: Catholic Relief Services and Bank Information Center.

Geef, P. 2005. 'Presentation to South Africa National Energy Association'. Sandton, 21 June.

Ghosh, S. n.d. 'Climate Change and the Market Politics of Environment'. *The National Forum of Forest People and Forest Workers*. http://www.sinkswatch.org.

Global Environment Facility (GEF) Secretariat. 1998. 'Study of the GEF's Overall Performance'. 2 March.

GlobeScan. 2006. '30-country Poll Finds Worldwide Consensus That Climate Change is a Serious Problem'. http://www.globescan.com/news_archives/csr_climatechange.html.

Gopinath, D. 2003. 'Doubt of Africa'. *Institutional Investor Magazine*, May.

Gore, A. 2006. *An Inconvenient Truth*. Lawrence Bender Productions.

Gorz, A. 1967. *Strategy for Labour*. Boston: Beacon Press.

———. 1973. *Socialism and Revolution*. Garden City: Anchor Press.

Gosling, Melanie. 2005. 'Pebble Bed to Cost R25bn'. *Cape Times*, 15 August.

———. 2006. 'SA the Top Emitter of Carbon Dioxide in the World'. *Cape Times*, 4 December.

Greenberg, S. 2002. 'Eskom, Electricity Sector Restructuring and Service Delivery in South Africa'. Cape Town: Alternative Information and Development Centre, June.

Groenewald, Y. 2007. 'SA will Solve Climate Change'. *Mail & Guardian*, 16 March.

groundWork. 2005. 'Whose Energy Future? Big Oil against the People of Africa'. http://www.groundwork.org.za.

Haites, E. 2006. 'Presentation to York University's Colloquium on the Global South'. Margaree Consulting, 25 January.

Halpern, S. 1992. *United Nations Conference on Environment and Development: Process and Documentation*. Providence, RI: Academic Council for the United Nations System (ACUNS). http://www.ciesin.org/docs/008-585/unced-home.html.

Hardin, G. 1968. 'The Tragedy of the Commons'. *Science*: 162.

Hathaway, T. 2005. 'Grand Inga, Grand Illusions?'. *World Rivers Review*, 20.2 (April). http://www.irn.org/pubs/wrr/issues/WRR.V20.N2.pdf.

Horner, C.C. 2002. 'Controlling Hypocritical Authority: Gore's Expertise'. *National Review Online*, April 23. http://www.cei.org/gencon/019,02972.cfm.

Horton, L. n.d. 'Environmental Justice and the CDM in Durban'. Honours thesis, Dartmouth College, New Hampshire.

Hosken, G. and S. Adams. 2005. 'What is the Matter at Pelindaba?'. *Pretoria News*, 29 April.

I-Net Bridge. 2005. 'Manuel Gives the Green Light to PBMR'. 23 February.

Intergovernmental Panel on Climate Change (IPCC). 1995. 'A Report of the Intergovernmental Panel on Climate Change (IPCC): Second Assessment Synthesis of Scientific-Technical Information Relevant to Interpreting Article 2 of the UNFCCC'. http://www.ipcc. ch/pub/sarsyn. htm.

———. 2001. 'Third Assessment Report: Summary for Policymakers'. http://www.ipcc.ch/pub/spm22-01.pdf.

International Emissions Trading Association (IETA). 2001. 'Meeting the Kyoto Protocol Commitments Summary – Domestic Emissions Trading Schemes'. http://www.ieta.org.

International Energy Agency (IEA). 2000. 'CO_2 Emissions from Fuel Combustion, 1971–1998'. http://www.iea.org.

International Institute for Energy Conservation, Hagler Bailly and the Stockholm Environment Institute. 1997. 'Carbon Backcasting Study'. 13 June.

Jury, M. 2004. 'Presentation to Durban Declaration Group'. Richards Bay, 9 October.

Kill, J. 2008. 'Lessons from the European Emissions Trading Scheme'. SinksWatch. http://www.sinks watch.org/pubs/2007%2009%20Lessons%20from%20the%20European%20 Emissions%20 Trading%20 Scheme%20_2_.pdf.

Kirby, R. 2006. 'Alec in Wonderland'. *Mail & Guardian*, 25 August.

Kockott, F. 2008. 'Minister Slams Divisive Whites'. *Sunday Independent*, 17 August.

Kristoff, N. 2007. 'The Big Melt'. *New York Times*, 16 August.

The Kyoto Protocol. http://www.unfccc.int.

Leonard, A. 2006. 'Free Market Environmentalism'. *Salon*, 8 March. http://www.salon.com.

Leslie, G. 2000. 'Social Pricing of Electricity in Johannesburg'. Masters research report submitted to the Faculty of Management, University of the Witswatersrand, Johannesburg.

Linden, G. 2006. 'Cloudy with a Chance of Chaos'. *FORTUNE*, 17 January. http://www.stopglobal warming.org/sgw_read.asp?id=1105281232006.

Lipow, G. 2006. 'Carbon Trading'. *PEN-L listserve*, 19 January.

Lohmann, L. 2001. 'Democracy or Carbocracy? Intellectual Corruption and the Future of the Climate Change Debate'. http://www.thecornerhouse.org.uk/item.shtml?x=51982%20.

———. 2002. 'The Kyoto Protocol: Neocolonialism and Fraud'. Talk given at 'Resistance is Fertile' gathering, The Hague, April.

———. 2005. 'Marketing and Making Carbon Dumps: Commodification, Calculation and Counter-factuals in Climate Change Mitigation'. *Science as Culture*, 14 (3).

———. ed. 2006. 'Carbon Trading: A Critical Conversation on Climate Change, Privatisation and Power'. *Development Dialogue*, 48 (Special Issue), September.

Mahlangu, L. 2005. 'Most South Africans Receive Free Water, Electricity'. SAPA, 17 March.

Martinez-Alier, J. 1998. 'Ecological Debt – External Debt'. Quito: Acción Ecológica. http://www.cosmovisiones.com/DeudaEcologica/a_alier01in.html.

———. 2003. 'Marxism, Social Metabolism and Ecologically Unequal Exchange'. Paper presented at Lund University Conference on World Systems Theory and the Environment, 19–22 September.

———. 2007. 'Keep Oil in the Ground: Yasuni in Ecuador'. *Economic and Political Weekly*, 20 October.

McCully, P. 2005. 'Comments on the World Bank's PCF/CDM Project Design Document for the China Xiaogushan Hydropower Project'. International Rivers Network, submitted to JCI CDM Center, 21 August.

McDonald, D. 2002. 'The Bell Tolls for Thee: Cost Recovery, Cutoffs and the Affordability of Municipal Services in South Africa'. Municipal Services Project Special Report. http://qsilver.queensu.ca/~mspadmin/pages/Project_Publications/Reports/bell. htm.

McGarr, P. 2005. 'Capitalism and Climate Change'. *International Socialism*: 107. http://www.isj.org.uk/index. php4? id=119&issue=107.

Monbiot, G. 2006. 'The Trade in "Carbon Offsets" is Based on Bogus Accounting'. *The Guardian*, 17 January.

———. 2007a. 'The Climate Talks are a Stitch-up, as No One is Talking about Supply'. *The Guardian*, 11 December.

———. 2007b. 'We've Been Suckered Again by the US. So Far the Bali Deal is Worse than Kyoto'. *The Guardian*, 17 December.

National Electricity Regulator. 2001. 'Annual Report 2000/01'. Johannesburg.

The National Institute of Public Health (RIVM). 2001. 'Health and the Environment'. http://www.rivm.nl.

Netto, M. and K.U. Barani Schmidt. 2005. 'CDM Project Cycle and the Role of the UNFCCC Secretariat'. In *Legal Aspects of Implementing the Kyoto Protocol Mechanisms*, eds. D. Freestone and C. Streck. Oxford: Oxford University Press.

NGO Alternative Treaties. 1992. 'Global Forum at Rio'. 1–15 June. http://habitat.igc.org/treaties/.

Olukoya, S. 2001. 'Environmental Justice from the Niger Delta to the World Conference Against Racism'. *CorpWatch*, Special Edition, 30 August. http://www.corpwatch.org/article.php?id=18.

Parikh, J.K. 1995. 'Joint Implementation and the North and South Cooperation for Climate Change'. *International Environmental Affairs*, 7 (1).

Plaut, M. 2004. 'US to Increase African Military Presence'. 23 March. http://www.bbc.co.uk.

Pressly, D. 2006. 'Erwin Backtracks on Koeberg "Sabotage" '. http://www.mg.co.za/articlePage.aspx?articleid=265860&area=/breaking_news/breaking_news_national/, 3 March.

Prototype Carbon Fund (PCF). 2002. 'Validation of the Plantar Project'. Report, 12 June.

———. 2003. 'Durban, South Africa: Landfill Gas to Electricity'. Project design document, final draft. Washington, DC: World Bank, January.

———. 2004a. 'Annual Report'. http://www.prototypecarbonfund.org.

———. 2004b. 'Durban, South Africa: Landfill Gas to Electricity'. Project design document. Washington, DC: World Bank, July. http://carbonfinance.org/pcf/Router. cfm?Page= Projects&ProjectID=3132# DocsList.

———. 2005a. 'Annual Report'. Washington, DC: World Bank Group. http://www. prototypecarbon fund.org.

———. 2005b. 'Carbon Market Trends 2006'. Washington, DC: World Bank Group.

Reddy, T. 2005a. 'Durban's Perfume Rods, Plastic Covers and Sweet-smelling Toxic Dump'. Durban: Centre for Civil Society Research Reports.

———. 2005b. 'Facing a Double Challenge'. Durban: Centre for Civil Society, University of KwaZulu-Natal. http://www.carbontradewatch.org.

———. 2006. 'Blinded by the Light.' *New Internationalist*, June.

The Renewable Energy and Energy Efficiency Partnership. n.d. 'CDM Housing Project to Become Replicable Energy Savings Model for South Africa'.

The Republic of Uganda (The Forest Department). 1998. 'Country Report on Assessment of the Intergovernment Panel on Forest Proposals'. June.

Robbins, T. 2002. 'Durban Signs SA's First Carbon Finance Deal'. *Business Day*, 13 November.

Rossouw, M. 2008. 'Rich Nations Play Divide and Rule'. *Mail & Guardian*, 2 September.

Rowell, A. 2001. 'Corporations Get Engaged to the Environmental Movement'. *PR Watch*, 8 (3).

Sandborn, C., W.J. Andrews and B. Wylynko. 1992. 'Preventing Toxic Pollution: Toward a British Columbia Strategy'. Vancouver, BC: West Coast Environmental Law Research Foundation.

Sasol. 2005. 'Project Identification Note: Sasol Natural Gas Conversion Project'. Submitted to the Designated National Authority, South Africa, 31 January.

———. 2006. 'Annual Financial Results 2006'. http://www.sasol.com.

Scheelhaase, J. 2001. 'International Greenhouse Gas Trading – New Business Options for Banks and Brokerage Houses'. Deutsche Bank Research, 7 December. http://www. dbresearch.com.

Schiermeier, Q. 2006. 'Methane Finding Baffles Scientists'. *Nature*, 12 January.

Schmidt, M. 2004. 'US Offers to Train and Equip Battalions'. *This Day*, 19 July.

Selva, M. 2006. 'Western Corporations are Exploiting Legal Loopholes to Dump Their Waste in Africa'. *The Independent*, 21 September.

Sole, S. 2007. 'Dangerous, covert liasons'. *Mail & Guardian*, 21 December.

South African Climate Action Network (SACAN). 2002. 'Can We Justify Selling Africa's Atmosphere?'. *Climate Action News*, July: 1.

South African Press Association (SAPA). 2004a. 'Denel to Benefit from US Defence Trade'. 21 July.

———. 2004b. 'Hot Interest in SA Nuclear Reactor'. 4 October. http://www.sinkswatch.org.

———. 2005. 'State Dismisses Nuclear Threat'. 28 April. http://www.sinkswatch.org.

SouthSouthNorth (SSN). n.d. 'Bellville South Landfill Gas Recovery and Use Project'. http://southsouthnorth.org/country_project_details.asp?country_id=5&project_id=72& project_type=1.

———. n.d. 'Kuyasa Low-Cost Urban Housing Energy Upgrade Project, Khayelitsha Cape Town, South Africa'.

Spalding-Fecher, A. 2000. 'The Sustainable Energy Watch Indicators 2001'. Cape Town: Energy for Development Research Centre, University of Cape Town, November.

Statistics South Africa. 2001. 'South Africa in Transition: Selected Findings from the October Household Survey of 1999 and Changes that have Occurred between 1995 and 1999'. Eds. R. Hirschowitz, W. Sekwati, D. Budlender. Pretoria: Statistics South Africa. http://www.info.gov.za/otherdocs/2001/survey.pdf.

Stern, Nicholas. 2006. *The Stern Review: The Economics of Climate Change*. Cambridge: Cambridge University Press.

Sustainable Energy and Economy Network (SEEN). 2007a. 'Enron's Pawns'. http://www. seen.org/PDFs/pawns.pdf.

———. 2007b. 'The Energy Tug of War'. http://www.seen.org/PDFs/Tug_of_war.pdf.

Sutcliffe, M. 2003. 'South Africa Cannot Afford to Waste Energy'. *The Mercury*, 27 February.

Taylor, I. 2003. 'Conflict in Central Africa: Clandestine Networks and Regional/Global Configurations, Review'. *African Political Economy*, 95: 49.

Toshiyuki, R., M.E. Drury, J. Belliveau, S. Kuhn and S. Bansal. 1999. 'Pollution Trading and Environmental Injustice: Los Angeles' Failed Experiment in Air Quality Policy'. *Duke Environmental Law and Policy Forum*.

Transnational Institute (TNI)/Foundation for Advancement in Science and Education (FASE). 2003. 'Where the Trees are a Desert: Stories from the Ground'. http://www.tni.org/ctw.

Tyler, E. 2006. 'CDM for Small, Sustainable Projects: Where is the Value Added?'. Ecosystem Marketplace Katoomba Group, 7 February. http://www.ecosystemmarketplace.com.

United Nations Development Programme (UNDP). 2004. 'South Africa Human Development Report 2003'. Pretoria.

United States Congressional Budget Office. 2008. 'Policy Options for Reducing CO_2 Emissions', February.

Vallette, J., D. Wysham and N. Martinez. 2004. *Wrong Turn from Rio: The World Bank's Road to Climate Catastrophe*. Washington, DC: Institute for Policy Studies.

Van Schalkwyk, M. 2006a. 'Speech by the Minsiter of Environmental Affairs and Tourism on the Occasion of the Opening of the Final Lead Authors Meeting of Working Group 2 of the Intergovernmental Panel on Climate Change's Fourth Assessment Report'. Somerset West, 11 September.

———. 2006b. 'Time in Sun for African Priorities'. *Business Day*, 22 November.

———. 2007a. 'Minister of Environmental Affairs and Tourism M. van Schalkwyk Spells out SA's "Climate Roadmap" for 2007 and Beyond'. Pretoria, 14 March.

———. 2007b. 'Keynote Address'. International Emissions Trading Association Forum, Washington, DC, 26 September.

———. 2007c. 'Statement by Marthinus Van Schalkwyk, Minister of Environmental Affairs and Tourism on Early Breakthrough on New Fund to Assist Developing Countries to Adapt to Climate Change'. Bali, Indonesia. http://www.info.gov.za/speeches/2007/07121109151001.htm.

Vedantam, S. 2005. 'Kyoto Credits System Aids the Rich, Some Say'. *The Washington Post*, 12 March.

White, C., O. Crankshaw, T. Mafokoane and H. Meintjes. 1998. *Social Determinants of Energy Use in Low Income Households in Gauteng*. Pretoria: Department of Mineral and Energy Affairs.

Williams, L. 2004. 'SA to Export Arms?'. *Business Day*, 21 July.

Winkler, H. and J. Mavhungu. 2001. 'Green Power, Public Benefits and Electricity Industry Restructuring'. Report prepared for the Sustainable Energy and Climate Change Partnership. Cape Town: Energy for Development Research Centre, University of Cape Town.

World Bank. 1992. 'Guidelines for Environmental Assessment of Energy and Industry Projects'. Technical paper 154/1992. *Environmental Assessment Sourcebook*, Vol III. Washington, DC: The World Bank.

———. 1993. 'The World Bank's Role in the Electric Power Sector: Policies for Effective Institutional, Regulatory, and Financial Reform'. Washington, DC: The World Bank.

———. 1999. 'Fuel for Thought: An Environmental Strategy for the Energy Sector'. Washington, DC: The World Bank.

———. 2004. 'Carbon Finance at the World Bank'. http://www.worldbank.org.

———. 2005. 'Where is the Wealth of Nations? Measuring Capital for the 21st Century'. Washington, DC: The World Bank. http://siteresources.worldbank.org/INTEEI/214578-1110886258964/20748034/All.pdf.

———. 2006a. 'Global Economic Prospects'. http://www-wds.worldbank.org/external/default/WDSContentServer/IW3P/IB/2006/12/06/000112742_2006120615 5022/Rendered/PDF/381400 GEP2007.pdf.

———. 2006b. 'The Role of the World Bank in Carbon Finance: An Approach for Further Engagement'. Washington, DC: The World Bank, 24 May.

The World Conservation Union (IUCN). 2005. 'Dire Consequences if Global Warming Exceeds 2 Degrees Says IUCN'. 29 November. http://news.mongabay.com/2005/1129-iucn.html.

World Rainforest Movement (WRM). 2000. 'Uganda: Carbon Sinks and Norwegian CO_2lonialism'. London.

———. 2002. 'Evaluation Report of V&M Florestal Ltda. and Plantar S.A. Reflorestamentos'. London.

Worldwide Fund for Nature (WWF). 2002. 'Position Paper on the Directive Proposal on Greenhouse Gas Emission Trading, February'.

Wysham, D. 2005. 'A Carbon Rush at the World Bank: Foreign Policy in Focus'. http://www.fpif.org/papers/0502wbank. html.

Index